A NATURALIST'S

SNAKES
OF
AUSTRALIA

Scott Eipper and Tyese Eipper

JOHN BEAUFOY PUBLISHING

First published in the United Kingdom and Australia in 2019 by John Beaufoy Publishing Ltd
11 Blenheim Court, 316 Woodstock Road, Oxford OX2 7NS, U.K.
www.johnbeaufoy.com

Photo Credits
Front cover Pilbara Death Adder © T. Eipper; bottom row, left to right Australian Bockadam © A. Elliott;
Water Python © S. Eipper; Slaty Grey Snake © S. Eipper. **Back cover** Keelback © S. Eipper. **Title page** Tiger
Snake © T. Eipper.
Contents page Shark Bay Sea Snake © S. Eipper. **Main descriptions** Photographs are denoted by a page number
followed by t (top), b (bottom), l (left), c (centre) or r (right).

Allen Allison 113b, **Shane Black** 4,20t, 32, 39bl, 40b, 41b, 76b, 97b, 99t, 100b, 102, 115t, 119c, 121c & bl,
122tl & b, 125tl, 130t, **Clay Bryce** 72tl, 92, **Elliott Budd** 17b, **Brian Bush** 52, 54b, 59b, 63b, 64t,65t,71b,107b,
125t & br, 129t, 133,139b,146t, 154b, **Casey Cannon** 123b, **Hal Cogger** 84b, 86, 90t, 94b, 104t, 150t, **Mark
Cowan** 157, **Graham Edgar** 43br, 47t, 72b, 90b, **Cody Eipper** 23b, **Scott Eipper/ Nature 4 You** 5-7, 10-13, 17t,
18-19, 20b, 21b, 22, 23c, 24, 25t, 26t, 27-28, 29t, 30b, 31, 33, 34b, 35b, 36b, 37-38, 39br, 40t, 41t, 42b, 46bl,
49, 50t, 51, 55, 56, 57t, 58, 62, 64b, 66, 68b, 69b, 73b, 74-75, 76t, 77-79, 87, 89b, 95b, 96, 97t, 98, 100t, 101t,
103t, 104b, 105t, 106, 107t, 109-110, 111b, 112, 113t, 114, 116,118, 119t, 120,121br, 122tr, 123t, 124, 126t,
127b,128b, 130b, 131t, 134t, 138, 141b, 142, 144b, 148b, 149b, 150 br, 151t, 152, 153b, 155b, 156t, 158, 161,
162t, 163t, **Tie Eipper/ Nature 4 You** 8, 91, 99b, **Adam Elliott** 23t, 26t, 35t, 36t, 42t, 50b, 68b,70, 111t, 129,
145t, 146b, **Ryan Ellis** 159b, **Jon Paul Emery** 48t, **Jules Farquhar** 135, **Ryan Francis** 54t, 63tr, 136bl, 137b,
150bl, **Glen Gaikhorst** 151b, **Prathamesh Ghadekar** 81b, **Richie Gilbert** 61b, 67t, **Philip Griffin** 148t, **Chay
Hoon** 44t, **Paul Horner** 53b, 80t, 82, 83t, 128t, 131b, 134b, 143b, 147t, 153t, **David Hunter** 163b, **Jordan
De Jong** 29b, 83b, 93b, **Daniel Kwok** 94t, **Jacob Loyacano** 85b, **Mike Lyons** 162b, **Phil Mangion** 61t, **Brad
Maryan** 44b, 84t, 108t, 117b, 154t, **Ross McGibbon** 108b, 117t, 126b, 155t, **Jake Meney** 25b, 30t, 53t 67b,
101b, **Angus McNab** 34t, **Jacob Nielsen** 45b, **Mark O'Shea** 45t, **Arne Rasmussen** 80b, 81t, **David Robinson**
46br, **Peter Rowland /Kape Images** 88t, **Mark Sanders/ Ecosmart Ecology** 60t, 73t, **Shawn Scott** 48b, 69b,
119b **Glenn Shea** 145b, **Ruchira Somaweera** 43t, **Gary Stephenson** 60b, **Klaus Stiefel** 72tr, **Peter Street** 89t,
Jason Sulda 21t, **Mike Swan** 149t, **Steve Swanson** 43bl, 46t, 59t, 71t, 85t, 105b, 132t, 139t, 141t, 147b, 159t,
Steve Tuckey 103b, **Kanishka Ukuwela** 95t, **Eric Vanderduys** 140, **Freek Vonk** 132b, **Lauren Vonnahme** 93t,
Harold Voris 88t, **Steve Wilson** 65b, 137b, 156t, 160b, **Justin Wright** 57b, 160t & **Anders Zimny** 39t, 47b,
63tl,115b, 136t & br, 143t, 144t

ISBN 978-1-912081-50-9

Edited by Krystyna Mayer
Designed by Gulmohur Press, New Delhi
Printed and bound in Malaysia by Times Offset (M) Sdn. Bhd.

DEDICATION
This book is dedicated to our sons Bailey and Cody. We thank you for your love, endless support and the
uncanny ability to keep us on our toes. A skill which has definitely helped us in the field. We love you boys.

·Contents·

INTRODUCTION

Australia is currently home to 216 species and 15 subspecies of snake, 147 of which are endemic. Five additional species of sea snake have been recorded adjacent to Australian waters – because it is likely that they may at times enter Australian territory they have been included in this book.

Many species are being investigated and it is inevitable that more will be described and resurrected from synonymy. Even during the course of writing this book, a new snake species was discovered in Queensland (see Cape York Bandy-bandy, p. 132). Australia is also home to two established introduced snakes, the Common Wolf Snake *Lycodon capucinus* and Flowerpot Snake *Indotyphlops braminus*. A third species, the Corn Snake *Pantherophis guttatus*, has been recorded in multiple states, but whether it has established viable populations in any areas is yet to be determined. It does, however, seem that this introduced species is established, due to the frequency at which they are found in parts of Sydney.

This book focuses on the classification of reptiles below the taxonomic rank of order (Squamata), in the suborder Serpentes – more commonly known as the snakes. Snakes are further divided worldwide into 25 families. Australia is home to seven snake families, including the file snakes, pythons and elapids, which encompass sea snakes and sea kraits, and blind snakes. The snakes known as colubrids were formerly part of a much larger family that has subsequently been divided – in Australia three ways into the colubrids, natracids and homalopsids. A brief overview of each family is given below.

File snakes are an old snake lineage comprising three species, two which are found in Australia. They are exclusively aquatic, using their amazing rough scalation to hold on to fish, which they constrict before swallowing. They are completely harmless to humans.

Pythons are thought to have evolved in either Australia or Indonesia and are most diverse in the region. They are some of the world's best-known snakes and include some truly massive species, such as the Burmese, Scrub, Reticulated and African Rock Pythons, all of which, despite their lack of venom, have used their amazing strength to cause the death of humans. This, however, is incredibly rare. Most pythons never attempt to harm humans unless provoked, and live alongside people in urban centres such as Sydney, Brisbane, Cairns, Darwin and Perth, where they provide pest control for rodents. All Australian species have heat-sensitive pits (in two species the pits are hidden behind the labial and rostral scales), which are used to locate prey. The pits are sensitive to about 0.003 of a degree Celsius change, which allows a python to sense the presence of warm-blooded prey. Many Australian pythons, such as carpet pytons and the Green Python, are among the world's commonly kept pets.

Scrub Python eating a meal

The **Colubridae** were thought to be the most diverse group of advanced snakes that were not vipers or elapids. Subsequent revisions have broken this family into many smaller families and subfamilies. It is thought that the three families represented in Australia arrived relatively recently from Southeast Asia via land bridges that were exposed during recent ice ages, or that they rafted across. This could explain the restriction of the colubrids to northern and eastern Australia, and the fact that most of the genera found in Australia are further diversified in other parts of Southeast Asia.

In Australia, the colubrids have been split into three families – the Colubridae, Natricidae and Homalopsidae. The **Colubridae** comprises the solid-toothed snakes and rear-fanged snakes of the genera *Boiga, Dendrelaphis and Stegonotus*, as well as the exotic introduced genera *Lycodon* and *Pantherophis*. The **Natricidae** are represented in Australia only by the Keelback *Tropidonophis mairii*. The **Homalopsidae** are represented by the genera *Cerberus, Fordonia, Myron* and

Pseudoferania. Until recently it was thought that solid-toothed snakes such as colubrids were completely harmless, without venom or a method of injection. This is incorrect, at least in many species. While some of the solid-toothed colubrids in Australia do not have fangs, they have saliva that contains various toxins such as anticoagulants, which are introduced into a bite via chewing.

Common Tree Snake

The **Elapidae** are found on all continents except Antarctica and contain snakes that strike fear in some people and generate wonder in others. Sea snakes are regarded as elapids by some authorities; others keep them in their own families, the Hydrophiidae and Laticaudidae. Both the Australian terrestrial elapids, and the sea snakes and sea kraits, are in their own subfamilies, the Hydrophiinae, within the Elapidae. The elapids are the most species-rich group of snakes in Australia.

Red Bellied Black Snakes emerging from embryonic sacs

Blind snakes are one of the world's least-known snake groups. There are three reasons for this. They are fossorial and rarely seen on the surface; they are morphologically conservative, with small rudimentary eyes, smooth scales and similar body shapes; and they are hard to find, with some species occuping very deep underground caverns and rarely being encountered. There are about 260 described species worldwide, although according to current phylogenetic assessments this number could increase dramatically. Many of the Australian species are known from fewer than 10 individuals.

Robust Blind Snakes hatching from eggs

SNAKE OR LIZARD?

- Lizards usually have eyelids, while snakes have a fixed scale called a spectacle or brille covering the eye. The exception to this are geckos and legless lizards.
- Lizards usually have ear openings, while snakes do not have ears and are fundamentally deaf to airborne sound.
- Most lizards have a broad, fleshy tongue. The exception to this are the varanoid lizards, such as monitors.
- Snakes do not have legs. Pythons and blind snakes have a vestigial pelvic girdle; in pythons this is shown on the surface by the cloacal spurs. Most lizards have limbs, but in some fossorial species and legless lizards these have become vestigial or non-existent.

Verreaux's Skink

Burton's Legless Lizard

Common Scalyfoot

Avoiding Snake Encounters and Snake-bite Prevention

- Never attempt to catch a snake.
- Never attempt to kill a snake – doing this is not only usually illegal, but will also put you at serious risk (20 per cent of all fatalities from snake bite occur as a result of the victim trying to handle or kill a snake).
- Never place your hands and feet where you cannot see them.
- Never walk around at night without a torch.
- Always wear shoes and long trousers in locations where snakes may be present.
- Keep rodents away from houses.
- Keep birds away from houses – their seeds attracts rodents, which in turn attract snakes.
- Take care when turning over potential shelter sites such as corrugated iron sheets, rocks and logs, as well as other debris.
- Store shoes indoors so nothing can seek out shelter inside them.
- Install and maintain screens on windows and doors.

Alway wear appropriate footwear in the bush

Snake Myths and Fallacies

- **Snakes are not repelled by vibration** Vibrating snake repellers are completely ineffective.
- **Venomous snakes and pythons cannot interbreed and form 'super snakes'.**
- **Not only harmless snakes climb** Many venomous snakes are found off the ground and climb well, particularly while hunting prey.
- **Snakes do not bite their tails and form hoops to go down hills**.
- **Snakes do not chase people** If they have perceived a threat they will defend themselves, but if given the opportunity they will almost always flee.
- **Snakes cannot outrun humans** The fastest snakes in the world can move at about 12km an hour in *short* bursts, while an average person is able to run at 20km per hour.
- **Snakes are not attracted to milk.**
- **Snakes will cross over horsehair ropes.**
- **Flowering plants do not repel snakes.**
- **Sea snakes can bite** Just like their terrestrial relatives, all sea snakes are capable of biting people.
- **Australia has the 'top ten' most toxic snakes in the world** Inland Taipans *Oxyuranus microlepidotus* are the most toxic species of snake to mice. Most quoted lists consist of Australian snakes only, with three exotic species as reference points.

SNAKE BITES AND FIRST AID

Note: this information is a guide for Australian snake bites only and is correct at the time of printing; all medical advice should be sought from professional medical staff.

If someone is bitten by a snake, early application of the correct first-aid practices can greatly increase the chances of them making a full recovery. First-aid measures are constantly advancing and improving, so it is important to stay current with the latest practices.

- *Seek urgent medical help in* Australia by *calling triple zero* **'000'**.
- **Do not** wash the bite site.
- Apply a pressure immobilization bandage and splint if the bite is on a limb, or a pressure pad if it is to the head, neck or torso, and minimize movement.
- **Do not** catch, chase or kill the snake – this involves extra movement (if done by the victim, increasing blood flow) and could result in further bites.
- **Do not** take alcohol, tea, stimulants, food or medication without expert advice.

IN THE FIRST INSTANCE
The **DRSABCD** action plan should be followed:

D	Look for Danger
R	Check for Response
S	Send for help
A	Clear the Airway
B	Sustain Breathing
C	Start CPR (if required)
D	Apply a Defibrillator if indicated

(DRSABCD is vital if a person has collapsed and is unresponsive)

Bite from a snake

Completed immobilized limb

Step 1 Wrap bandage over bite site

Step 2 Extend bandage up entire limb

Step 3 Mark site with time of incident

PRESSURE IMMOBILIZATION BANDAGING (PIB) FIRST AID

There are two components that must be satisfied – pressure over the bitten limb and focal plus general immobilization. This involves the application of:
1. A broad (minimum 75mm wide) elastic bandage to the entire bitten limb at a very firm pressure of at least 40mmHg for an arm and 55mmHg for a leg. The Australian Venom Research Unit (AVRU) recommends SETOPRESS TM High Compression Bandages as these bandages relax very little with prolonged application.
2. Splints to effectively immobilize the entire limb, in combination with laying the victim down and completely still to minimize any movement. Do not use a sling.

ALWAYS SEEK MEDICAL ATTENTION FOLLOWING A SNAKE BITE

Move away from the area where the bite occurred (if required), lie the patient down and keep them calm. Any movement of the limb quickly results in venom absorption and must be prevented; therefore first aid must be an immediate priority after a snake bite.

Do not allow the patient to walk. In the case of a snake bite to a lower limb, splinting of both legs should be carried out to completely immobilize the lower half of the body.

In rare cases, a person may be bitten on the body, face or neck. In these cases direct pressure should be applied over the bite site with a pressure pad made from a cloth (a hand towel, T-shirt, or similar.) and held firmly in place until medical attention can be sought.

RESPIRATORY INSUFFICIENCY MANAGEMENT

The bite from any venomous snake has the potential to cause envenomations that may result in an allergic reaction. In its most severe instance, this can lead to difficulty in breathing, which can be a result of anaphylaxis. Symptoms include acute, rapid onset illness, tingling around the mouth, swelling of the lips, tongue and face, tightness in the throat, difficult or noisy breathing, rash and hives, difficulty talking, coughing, dizziness, vomiting and abdominal pain.

If **anaphylaxis** is suspected:
* *Seek urgent medical help in Australia by calling triple zero '000'.*
* Lay the person flat on their back (or seated if breathing is difficult).
* Obtain and follow the instructions on an epinephrine (adrenaline) pen – usually injected in the outer thigh.
* **Do not** allow the victim to stand or walk around.
* Monitor breathing and commence CPR (30 chest compressions followed by two rescue breaths) if required, and continue until the breathing is normal and stable.
* Monitor the victim for at least four hours.

Habitats

Australian snakes are found in a wide variety of habitats. Some species are generalists, while others are restricted to small microhabitats within larger ecological communities. A sample of the variety of important habitats to Australian herpetofauna is provided below.

Alpine herb fields

Sand ridges with spinifex

Black-soil plains

Brigalow open woodland

Coral reefs

Mallee woodland

Mangroves

Open woodland

Rocky plains

Chenopod scrubland

Rocky spinifex woodland

Rocky escarpments

Rainforests

Forest creeklines

Floodplains

IDENTIFICATION

Pointed snout

Pointed snout

Strongly trilobed snout

Bluntly angular snout

Rounded snout

Rounded snout

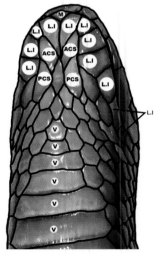

Head scalation (ventral)

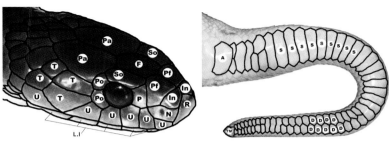

Elapid head scalation   *Tail scalation*

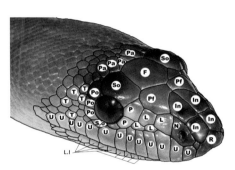

Python scalation

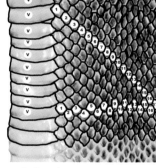

Midbody scalation

KEY TO DIAGRAMS

HEAD SCALATION

ACS Anterior chin shield (pregenial); **F** Frontal; **In** Internasal; **LL** Lower labial (infralabial); **L** Loreal; **M** Mental; **N** Nasal; **P** Preocular; **Pa** Parietal; **PCS** Posterior chin shield (postgenial); **Pf** Prefrontal; **Po** Postocular; **R** Rostral; **SO** Supraocular; **S.o** Subocular; **T** Temporal; **UL** Upper labial (supralabial).

BODY SCALATION

A Anal; **D** Divided subcaudal; **S** Single subcaudal; **Tm** Terminal; **V** Ventral.

Using This Book

This book is designed to provide up-to-date information on each of Australia's snakes. It is meant as an introductory guide for assisting in identification. In many cases, key points of reference to separate a species from its nearest relatives are provided. The identification of many species relies on the use of scale counts, distribution, scale conditions and sometimes internal organ position. Taxonomy follows Cogger 2018 with the exception of the addition of newly described taxa. Both species and subspecies are given their own entries. Subspecies are clearly identified by the additional subspecific name in the entry's heading.

DISTRIBUTION KEY

GDR Great Dividing Range; **NSW** New South Wales; **NT** Northern Territory; **PNG** Papua New Guinea; **Qld** Queensland; **SA** South Australia; **Tas** Tasmania; **Vic** Victoria; **WA** Western Australia.

POTENTIAL DANGER RANKINGS

DANGEROUSLY VENOMOUS Venom can kill a healthy person (adult or child).
VENOMOUS Venom is unlikely to kill a healthy person, but may cause serious systemic effects.
POTENTIALLY DANGEROUS Could kill a person in unusual circumstances.
HARMFUL Non-lethal venom, except to people who are at risk of severe allergic reactions and anaphylaxis; may cause a bite that could require stitches.
HARMLESS Bites may break the skin but unlikely to or unable to cause harm to a healthy person.

KEY FEATURES AND MEASUREMENTS

Sizes quoted in species descriptions for body measurements (where available) are average maximum sizes, but exceptions can occur. Breeding information (such as clutch/litter sizes), is taken from current literature and should be treated as an indicative value, as ongoing research can change the values provided.

DSR Dorsal scale rows; **MB** Midbody scale rows range; **SUB** Subcaudal scales/pairs range; **TL** Total length; **VENT** Ventral scale count range.

GLOSSARY

anal scale/s Scale or scales that are anterior to the cloaca (see diagram, p. 13).

anaphylaxis Severe, fast-acting allergic reaction that can be fatal.

anterior Front of body; towards front.

anticoagulant Substance that prevents clotting of blood.

antivenom Medication used to treat venomous bites and stings.

aquatic Living in or near water.

arboreal Living predominantly above the ground, that is in vegetation or trees.

basking Act of a snake exposing itself to increased temperature in order to raise its core body temperature.

cathemeral Active at any time of day or night.

caudal At or near rear half of body.

chenopod shrubland Semi-arid plains vegetated with saltbush, samphire and similar.

clutch Number of eggs laid by female snake in a single reproductive event.

coagulant Component of venom that causes blood to clot.

constriction In snakes, act of coiling tightly around an animal, causing suffocation.

crepuscular Active at dawn and dusk.

cryptic Disguising appearance, through either colour and pattern, or habits.

diurnal Active during daytime.

dorsal Of, on or relating to upper half or top of a structure or body.

family Taxonomic rank above genus and below order.

fangs Sharp, grooved or hollow teeth modified to pierce living tissue and inject venom.

feral animal Introduced animal that has become established in the wild.

fossorial Living or active beneath the soil surface.

gene Basic unit of genetic control. Each gene has a specific function and is found on a specific section of a specific chromosome.

genus (*pl.* **genera**) Taxonomic group above species and below family.

gravid Pregnant; abdominal cavity contains formed eggs or young.

haemolytic/haemolysin Components of venom that cause red blood cells to be destroyed.

heliothermic Refers to individual that must bask to raise its body temperature.

herpetofauna Collective term referring to group of amphibians and reptiles.

hybrid Genetic combination as a result of the mating of two different species or subspecies.

intraspecific In reference to a process or state within a species or subspecies.

juvenile Young individual.

litter Number of young born by female snake in a single reproductive event.

maxilla Jaw (upper).

midbody scale rows Pertains to diagonal line of scales counted from a ventral scale over body to ventral scale on other side (see diagram, p. 13).

morphological Pertaining to form or structure of an animal, especially its external appearance.

myotoxins Components of venom that attack the muscle tissue.

nape Back of neck.

nasal scale The scale that contains nostril (see diagram, p. 13).

necrosis Premature death of cells in living tissue.

neonate Newborn animal up to the age of six weeks.

neurotoxin Component of venom that attacks nervous system.

nocturnal Active during the night.

nuchal Area where head and neck join.

order Taxonomic rank above family and below class.

oviparous Reproducing by laying eggs.

ovoviviparous Reproducing by forming unshelled egg sacs to house developing young, which are held inside female until ready to hatch, then expelled either still within egg sac or after leaving it.

pelagic Occurring in the open ocean.

poison Substance that is harmful once ingested, inhaled or absorbed through skin.

posterior Rear of body; towards rear.

ptosis Uncontrollable drooping of eyelids. Often a sign of systemic neurotoxic envenomation.

race An intraspecific category below a subspecies that is characterised by a particular trait or property.

rostral Scale on end of snout.

species Basic unit of taxonomic classification.

species complex Group of animals that comprise both described and undescribed taxa that are currently lumped under a single species name.

stippling Patterns or markings created by grouping together of numerous small dots.

subcaudal scales The scales that are posterior to the cloaca (see diagram, p. 13).

subspecies Taxonomic category that is a variation in a primary (nominate) species brought about by geographical or genetic isolation, usually characterized by a variation in morphological or genetic features.

supralabial scales The scales only on side of head running along upper lip (see diagram, p. 13).

sympatry Refers to individuals that share the same geographical area.

taxonomy Study of plants and animals leading to their description, classification and naming.

terrestrial Living on or near the ground surface.

torpor Dormant period of physical inactivity usually characterized by reduced body temperature and slowed metabolic rate.

trilobed In reference to the shape of some blind snakes' heads viewed from above. See image p. 12.

troglodyte A cave-dwelling species.

venom Toxin injected by an animal to subdue prey or for defence against predators.

ventral Undersurface or belly of an animal.

ventral scales Enlarged scales running along underside of a snake (see diagram, p. 13).

vertebral Along line of spine.

vertebrates Animals that have a backbone.

vestigial Refers to remnant of appendage or other structure that has lost its original purpose through evolution.

viviparous Reproducing by giving birth to live young.

waif Individual that is transported by natural events such as storms, currents or floods into waters or regions outside the typical range for the species.

ACROCHORDIDAE (FILE SNAKES)
This small family contains three species, two found in Australia and a third in southern Asia. Entirely aquatic, they are graceful and agile when submerged, but almost helpless if stranded on land. Outside Australia they are often referred to as 'wart snakes' due to their distinctive scalation.

Arafura File Snake ■ *Acrochordus arafurae* TL 220cm

DESCRIPTION Upperparts grey to brown or blackish, marked with darker pigment forming cross-bands and spotting. Underside grey. Underside forms keel while swimming. Markings more defined on juveniles and subadults than on adults. Midbody scales conical in shape, an adaption thought to be used for holding prey during constriction. MB 120–180 rows. **DISTRIBUTION** Northern Qld, west of GDR, to WA border. Also southern PNG. **HABITAT AND HABITS** Completely aquatic. Lives in freshwater rivers, streams, pools and billabongs.

Nocturnal. Ambush predator, anchoring itself among submerged tree roots, and preying on fish. Occasionally found in estuarine environments, where it is vagrant. Gives birth to 1–127 young. HARMLESS

Little File Snake ■ *Acrochordus granulatus* TL 120cm

DESCRIPTION Upperparts grey to brown or blackish, marked with lighter pigment forming pale cross-bands. Underside grey. Underside forms keel while swimming. Markings more defined on juveniles and subadults than on adults. Midbody scales conical in shape, an adaption thought to be used for holding prey during constriction. MB 90–160 rows.
DISTRIBUTION Coastal regions from Townsville, Qld, north and west to Kalumburu, WA. Also waters surrounding PNG and Southeast Asia. **HABITAT AND HABITS** Completely aquatic. Lives in brackish and marine ecosystems. Occasionally enters freshwater zones of river systems. Nocturnal. Hides in crab burrows by day. Hunts primarily in intertidal zones, preying on fish, and occasionally eating small crabs and carrion. Gives birth to 6–12 young. HARMLESS

> **PYTHONIDAE (PYTHONS)**
> This family includes some of Australia's best-known snakes such as carpet and
> diamond pythons, and the beautiful Green Python of far northern Qld. It comprises
> both the world's and Australia's longest snakes. Australian species are diverse in their
> ecology and found over much of the country. Most have exposed labial pits used for
> detection of warm-blooded prey. All lay eggs.

Children's Python ■ *Antaresia childreni* TL 100cm

DESCRIPTION Reddish or yellowish-brown above, often with darker brown blotched
markings covering dorsum. Occasionally without markings. Lacks plain white stripe along

lower side of first third of body (see Stimson's
Python, p. 20). Some island populations
are melanistic. White to cream below. MB
37–49 rows, 255–300 VENT, SUB 30–45 a few
anteriorly single and remaining divided, anal
single. **DISTRIBUTION** North-western Qld
to Kimberley region, WA. **HABITAT AND
HABITS** Terrestrial, living in tropical savannah,
rock outcrops and woodland. Nocturnal. Hunts
for lizards, frogs and rodents. Lays 5–24 eggs.
Named after a museum worker rather than for
its suitability as a pet species. HARMLESS

Spotted Python ■ *Antaresia maculosa* TL 140cm

DESCRIPTION Dark or yellowish-brown above, with darker brown to black blotched
markings covering dorsum. Some island populations have lighter markings. White to

cream below. MB 35–40 rows, 245–290
VENT, SUB 30–45 mainly divided
(sometimes a few single anteriorly), anal
single. **DISTRIBUTION** Eastern Qld to
north-east NSW. Also Torres Strait islands
and PNG. **HABITAT AND HABITS**
Terrestrial, living in brigalow, rainforests,
rock outcrops and woodland. Predominantly
nocturnal. Hunts for lizards, frogs and
rodents. Often uses cave entrances as
ambush locations to capture bats. Lays 4–19
eggs. Known to hybridize with Stimson's
Python (see p. 20) where they overlap in
distribution. HARMLESS

Pygmy Python ■ *Antaresia perthensis* TL 80cm
(Anthill Python)

DESCRIPTION Reddish or occasionally yellowish-brown above, with darker flecking covering dorsum. Some populations almost patternless. White to cream below. MB 31–35 rows, 205–255 VENT, SUB 30–45 mainly divided (sometimes a few single anteriorly), anal single. **DISTRIBUTION** WA from Goldsworthy to Mt Magnet, including Pilbara, Gascoyne and Murchison regions. **HABITAT AND HABITS** Terrestrial, living in spinifex-dominated grassland with rock outcrops and rocky gorges. Nocturnal. Hunts for lizards and the occasional frog. Often shelters in termite mounds. Lays 2–12 eggs. The world's smallest python. HARMLESS

Western Stimson's Python ■ *Antaresia stimsoni stimsoni* TL 100cm

DESCRIPTION Reddish or occasionally yellowish-brown above, with darker spotting covering dorsum. Pale lateral line from neck extending along first third of body. White to cream below. MB 35–50 rows, 260–302 VENT, SUB 30–45 mainly divided (sometimes a few single anteriorly), anal single. Differs from Stimson's Python (see p. 20) by higher ventral count, position of eye above the 6th and 7th supralabials (vs 5th and 6th), and distribution. **DISTRIBUTION** WA from Perth to Kimberley, and east across to Kalgoorlie in south, and Fitzroy Crossing in north. **HABITAT AND HABITS** Terrestrial, living in grassland, heaths with rock outcrops, open woodland and rocky gorges. Nocturnal. Hunts for lizards, frogs and mammals. Lays 6–16 eggs. HARMLESS

Stimson's Python ■ *Antaresia stimsoni orientalis* TL 120cm

DESCRIPTION Reddish or occasionally yellowish-brown above, with darker spotting covering dorsum. Pale lateral line from neck extending along first third of body. White to cream below. MB 35–50 rows, 243–284 VENT, SUB 30–45 mainly divided (sometimes a few single anteriorly), anal single. Differs from Western Stimson's Python (see p. 19) by

lower ventral count, position of eye above the 5th and 6th supralabial (vs 6th and 7th) and distribution. **DISTRIBUTION** Eastern WA, across central Australia into arid SA, including northern Flinders Ranges, north-western NSW, NT and across most of arid Qld, to Einasleigh Uplands and Mt Carbine area. **HABITAT AND HABITS** Terrestrial, living in open woodland, spinifex-dominated grassland, black-soil plains and rocky gorges. Predominantly nocturnal. Hunts lizards, mammals and frogs. Lays 4–17 eggs. Known to hybridize with the Spotted Python (see p. 18). Mt Carbine animals may be a separate species, but further work is required to determine this. HARMLESS

'Mt Carbine' race

Black-headed Python ▪ *Aspidites melanocephalus* TL 300cm

DESCRIPTION Reddish to yellowish-brown or white above, with dark brown to black cross-bands. Head and neck black. Cream to yellow below with orange, brown and black markings. MB 50–65 rows, 315–359 VENT, SUB 60–75 all single, occasionally a few divided posteriorly, anal single. One of two pythons with labial pits covered by scales.

DISTRIBUTION From Mundubbera, Qld, across northern Australia to Exmouth, WA. **HABITAT AND HABITS** Found in open woodland, rainforest margins to spinifex dominated grassland, black-soil plains and rocky gorges. Terrestrial, usually on rocky or clay loam soils. Nocturnal, but occasionally basks by day. Hunts lizards, snakes and mammals. Lays 5–20 eggs. Bites, while completely harmless, can register on venom-detection kits, from residue of a snake's oral secretions. HARMLESS

Woma ▪ *Aspidites ramsayi* TL 220cm
(Sand Python)

DESCRIPTION Reddish-orange to yellowish-brown or grey above, with light to dark grey bands. May occasionally have purplish-black to black bands. Older individuals often fade with age. Head and neck yellow to orange. Sometimes dark marks over eyes. Orange to yellow below with or without dark markings. MB 43–65 rows, 273–315 VENT, SUB 40–55 all single, occasionally a few divided posteriorly, anal single. One of two pythons with labial pits covered by scales. **DISTRIBUTION** Two distinct populations: one from

Westmar, Qld, across Australia through western NSW, northern SA, south and central NT, to south of Broome, WA; the other in WA, from Shark Bay to Perth, into goldfields east of Kalgoorlie. **HABITAT AND HABITS** Found in open woodland, brigalow, deserts and mulga woodland. Terrestrial, and usually seen on sandy soils. Nocturnal. Occasionally hunts in trees for lizards and snakes. Will also eat mammals and birds. Lays 5–15 eggs. Bites, while completely harmless, can register on venom-detection kits, from residue of snake's oral secretions. South-western population listed by local government as rare and locally endangered. HARMLESS

Water Python ▪ *Liasis fuscus* TL 220cm
(Rainbow Serpent)

DESCRIPTION Greenish-brown to grey above.
Scales have a strong iridescence, hence the species'
alternate name of Rainbow Serpent. Underside
white beneath head, yellow to orange below
body and grey under tail. Lips usually peppered
with grey or black over white. MB 40–50 rows,
270–300 VENT, SUB 60–90 divided and anal
single. **DISTRIBUTION** From Conway, Qld, across
northern Australia to Broome, WA. **HABITAT
AND HABITS** Terrestrial to semi-aquatic, and
found in moist environments in tropical savannah
and woodland, as well as in swamps and floodplains.
Nocturnal but also basks by day. Hunts for rodents,
birds and occasionally reptiles including crocodiles.
Lays 9–20 eggs. Phylogenetic data suggests eastern
and western populations are different taxa, but this
requires further investigation. HARMLESS

Western population

Eastern population

Olive Python ▪ *Liasis olivaceus olivaceus* TL 400cm

DESCRIPTION Pale to dark brown above to grey. Belly cream to pale yellow. MB 60–80
rows, 321–377 VENT, SUB 96–119 all divided and anal single. **DISTRIBUTION** Across
northern Australia from along Selwyn Range, Qld, across to Kimberley, WA. **HABITAT
AND HABITS** Terrestrial and semi-arboreal, living in open woodland, savannah, swamps,
rocky hillsides and river edges. Occasionally found in houses. Nocturnal, occasionally
basking by day, particularly during cooler months of the year. Ambush predator that feeds
on mammals, reptiles and birds. Lays 5–12 eggs. HARMFUL

Pilbara Olive Python ■ *Liasis olivaceus barroni* TL 450cm

DESCRIPTION Dark brown above to grey. Belly cream to yellow. MB 58–63 rows, 374–411 VENT, SUB 90–110 all divided and anal single. There are unconfirmed reports that this species may reach up to 6.5m in length. However, none of the official records has demonstrated that the species reaches this length. **DISTRIBUTION** Only in Pilbara region and Mt Augustus, WA. **HABITAT AND HABITS** Terrestrial and semi-arboreal, living in open woodland, in rocky gorges usually with permanent water. Nocturnal.

Ambush predator that feeds on wallabies, rodents, flying foxes and birds. Young animals eat rodents and lizards. One recorded clutch of 8 eggs cut from dead wild individual. Phylogenetic data suggests it may be a different species and future work may elevate the taxon, but this requires further investigation. State-based assessments regard this subspecies as vulnerable. HARMFUL

Centralian Carpet Python ■ *Morelia bredli* TL 300cm
(Bredl's Python)

DESCRIPTION Reddish-brown to dark brown above, often darker on rear third, with white, yellow and black markings, transitioning into bands on rear third. White to cream below with orange-brown and black spotting. MB 50–55 rows, 280–310 VENT, SUB 85–95 all divided and anal single. **DISTRIBUTION** NT around MacDonnell Ranges across to Hart's Range.

HABITAT AND HABITS Both terrestrial and arboreal, living along gorges, and dry creek beds with old-growth eucalypts that have deep hollows. Occasionally enters houses in Alice Springs. Nocturnal. Feeds mainly on mammals and birds. Sometimes treated as subspecies of Carpet Python (see p. 28). Lays 15–45 eggs. HARMLESS

Rough-scaled Python ■ *Morelia carinata* TL 200cm

DESCRIPTION Dark brown or grey, with irregular white bands and streaks that form blotches and bands. Underside white with brown peppering towards rear. Marked shift in colour from day to night; individuals that are brown by day become silver at night. Midbody scales distinctly keeled. MB 45–65 rows, 298–292 VENT, SUB 83–89 mainly divided and anal single.

DISTRIBUTION Kimberley on Mitchell River Plateau and adjacent areas; also Bigge Island, WA. **HABITAT AND HABITS** Found in vine thickets in deep gorges. Nocturnal. Has very long teeth, presumably as an adaptation to its rodent prey, but will also eat birds. Lays 10–15 eggs. HARMLESS

Western Carpet Python ■ *Morelia imbricata* TL 270cm

DESCRIPTION Very variable. Brown above, with black, grey or yellow with white to cream or yellow markings that form irregular bands, stripes and blotches. Belly white with grey or black flecks and spots. Juveniles reddish. MB 41–49 rows, 239–276 VENT, SUB 63–82 mainly divided and anal single. **DISTRIBUTION** From Carrarang Station, across south-

west corner of WA, to Esperance. In SA on western Eyre Peninsula and on islands of St Francis group. **HABITAT AND HABITS** Both terrestrial and arboreal, living in forests, open woodland, rocky hills, grassland, mallee and heathland. Commonly found in houses. Nocturnal, but often basks by day. Eats mammals, birds and occasionally lizards. Lays 5–27 eggs. Once thought to be subspecies of Carpet Python (see p. 28); genetic and morphological evidence elevated it to a full species. HARMLESS

Inland Carpet Python ■ *Morelia spilota metcalfei* TL 220cm
(Murray Darling Carpet Python)

DESCRIPTION Very variable. Body brown, black, grey or yellow, with white, cream, orange, yellow and pale brown markings that form irregular bands, stripes and blotches. Belly white with grey or black flecks and spots. Can acquire a heavy build with age – large individuals routinely have a girth greater than that of a man's arm. Colouration shifts from dark in south of range to lighter with less black in north. Juveniles reddish. MB 40–65 rows, 240–310

VENT, SUB 60–95 mainly divided and anal single. **DISTRIBUTION** Along Murray Darling basin, west of GDR, from Pyramid Hill, Vic, west to eastern SA and through western and central NSW and Qld, to Dajarra in the north. **HABITAT AND HABITS** Both terrestrial and arboreal, living in open woodland, rocky hills, grassland, savannah, swamps and river edges. Commonly found in houses. Nocturnal, but often basks by day. Eats mammals, birds and occasionally lizards as adult, and lizards as juvenile. Lays 8–43 eggs. HARMLESS

Carpet Python ■ *Morelia spilota spilota* TL 320cm
(Diamond Python)

DESCRIPTION Very variable. Body brown, black, grey or yellow, with white to cream or yellow markings that form irregular bands, stripes and blotches. Southern animals black with yellow flecking. Belly white with grey or black flecks and spots. Attains heavy build with age; large individuals from northern NSW and southern Qld routinely have a girth greater than that of a man's arm. Colouration shifts from dark in south of range to lighter with less black further north. Juveniles reddish. MB 40–65 rows, 240–310 VENT, SUB 60–95 mainly divided and anal single. **DISTRIBUTION** Much of Australia from Marlo, Vic, to Kimberley region, WA. Also PNG. **HABITAT AND HABITS** Both terrestrial and arboreal, living in rainforests, forests, open woodland, rocky hills, grassland, savannah, swamps, mangroves and river edges. Commonly found in houses and other properties. Nocturnal, but often seen basking by day. Eats mammals, birds and occasionally lizards as adult. Lays 5–56 eggs. Genetic evidence and lack of reliable morphology has led previously accepted subspecies to be placed into synonymy and retained as geographic races. HARMFUL

'Jungle' race

'Top End' race

'Diamond' race

Carpet Python continued

'Cape York Savannah' race

'Coastal' race

Green Python ■ *Morelia viridis* TL 150cm

DESCRIPTION Bright green, usually with broken white vertebral stripe. Some individuals have a bluish flush. Juveniles yellow with brown broken vertebral stripe, and brown and white flecks. Tail-tip usually a different colour to body, and used for caudal luring of its prey. Underside yellow or white with green, blue and black markings. MB 50–70 rows, 210–260 VENT, SUB 90–110 mainly or all divided and anal single. **DISTRIBUTION** Qld on Cape York Peninsula, in Iron and McIlwraith Ranges. Also southern PNG. **HABITAT AND HABITS** Arboreal, living in rainforests and vine forests. Nocturnal, but often seen basking by day. Eats mammals, birds and occasionally lizards. Lays 5–26 eggs. HARMLESS

Juvenile

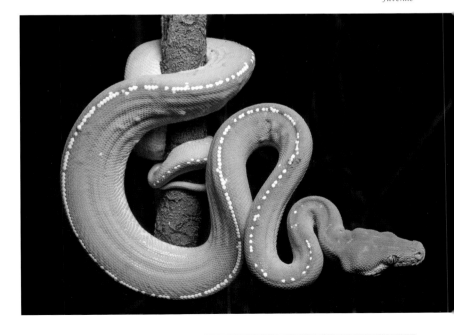

Scrub Python ■ *Simalia amethistina* TL 550cm
(Amethystine Python)

DESCRIPTION Top of body brown with yellowish and black markings forming irregular bands and blotches. Overall pearlescent sheen caused by refraction. Belly white to cream without markings. MB 35–50 rows, 270–340 VENT, SUB 80–120 mainly or all divided and anal single. Previously placed in the genus *Morelia*. **DISTRIBUTION** North-east Qld and PNG. **HABITAT AND HABITS** Lives in forests, savannah margins, swamps, mangroves and river edges. Commonly found in houses. Terrestrial but will occasionally climb trees.

Nocturnal, occasionally basking by day, particularly during cooler months of the year. Ambush predator – diet includes large prey such as wallabies, brush-turkeys, bandicoots, possums and occasionally lizards. Lays 5–12 eggs. Very powerful and when very large can overpower a healthy adult human. Australia's largest snake. Identity of species in Australia unclear – possibly a species complex, requiring further work. POTENTIALLY DANGEROUS

Oenpelli Python ■ *Simalia oenpelliensis* TL 500cm
(Narawan)

DESCRIPTION Dark brown or grey, with irregular white bands and streaks that form blotches and bands. Underside white with brown peppering towards rear. Marked shift in colour from day to night; individuals that are brown by day become silver at night. MB 65–75 rows, 420–450 VENT, SUB 150–170 mainly divided and anal single. **DISTRIBUTION**

Western Arnhem Land, NT. **HABITAT AND HABITS** Found around Kombalgie sandstone gorges and outcrops, associated woodland and creek lines. Often seen around rock crevices and in trees. Nocturnal. Feeds on mammals and birds. Lays 10–15 eggs. Previously placed in the genus *Morelia*. HARMFUL

COLUBRIDAE (COLUBRID SNAKES)

The colubrids are relative newcomers to Australia, as shown by the lack of species diversity and their northern/north-eastern distribution patterns. These solid-toothed species were thought to be non-venomous, but subsequent research has shown that all species have rudimentary toxins within the saliva. This, coupled with their elongated teeth, makes them technically venomous, although completely harmless barring an allergic reaction. Two species occur in Australian territories due to human introduction.

Brown Tree Snake ■ *Boiga irregularis* TL 200cm
(Night Tiger; Doll's Eye Snake)

DESCRIPTION Reddish-brown to dark brown above, with darker reticulations. Alternatively, reddish-orange with contrasting white bands. White, yellow or orange below. MB 19–23 rows, 225–265 VENT, SUB 85–130 all divided and anal single.

Distinct row of enlarged vertebral midbody scales. **DISTRIBUTION** Across northern and eastern Australia from Kimberley, WA, to Wollongong, NSW. Also PNG, Indonesia and introduced to Guam. **HABITAT AND HABITS** Arboreal, living in forests, gorges, savannah, swamps and mangroves. Commonly enters houses. Nocturnal, but basks by day. Feeds mainly on mammals and birds. Lays 3–11 eggs. Possibly a species complex. HARMFUL

'Night Tiger' race

Northern Tree Snake ■ *Dendrelaphis calligastra* TL 120cm

DESCRIPTION Grey to dark brown above, occasionally with black speckling. Anterior edge of most midbody scales pale, only becoming visible when snake inflates its body in response to a threat. On side of head is a bright yellow to white streak with upper margin edged with black and extending from snout beneath eye on to neck; this is not present in the similar Common Tree Snake (see opposite). Greyish and mottled with darker pigment below. MB 13–15 rows, 180–230 VENT, SUB 90–150 all divided and anal divided. **DISTRIBUTION** North-east Qld, from Mt Elliott on to Torres Strait Islands. Also PNG and Indonesia. **HABITAT AND HABITS** Arboreal, living in forests, savannah, swamps and mangroves. Occasionally found sleeping on vegetation at night. Diurnal. Feeds mainly on frogs and lizards. Lays 5–12 eggs. HARMLESS

Common Tree Snake ■ *Dendrelaphis punctulatus* TL 170cm
(Green Tree Snake; Yellow-bellied Black Snake; Common Bronzeback)

DESCRIPTION Very variable – colouration depends on location. A number of colour forms, including golden-yellow with grey or white head, golden-brown with darker head, black with lighter underside, blue to blue-grey, and green with yellow flush to head. Anterior edge of most midbody scales pale white to pale blue, only becoming visible when snake inflates its body in response to a threat. Underside whitish, blue, green, black or yellow. MB 13–15 rows, 180–230 VENT, SUB 100–150 all divided and anal divided. **DISTRIBUTION** Eastern NSW, from Batemans Bay, through eastern Qld, and across Top End, NT, and Kimberley region, WA. Also PNG and Indonesia. **HABITAT AND HABITS** Arboreal, living in forests, savannah, swamps and mangroves. Diurnal. Very common in some urban areas. Feeds mainly on frogs and lizards. Lays 5–14 eggs. HARMLESS

Common Wolf Snake ■ *Lycodon capucinus* TL 80cm

DESCRIPTION Grey to dark brown above, with white to yellow reticulations that form irregular cross-bands. Lower flanks lighter. Usually has pale nuchal band. White to pale grey below. MB 17 rows, 170–225 VENT, SUB 50–80 all divided and anal divided. **DISTRIBUTION** Introduced to Australia on Christmas Island. Throughout southern

Asia. **HABITAT AND HABITS** Terrestrial and arboreal, living in forests, savannah, swamps and mangroves. Commonly enters houses on Christmas Island. Nocturnal. Feeds on lizards. Lays 4–11 eggs. Accidental introduction to Christmas Island has been detrimental to native lizard fauna and has been implicated in local extinction in the wild of four species, and as threatening two others. HARMLESS

Corn Snake ■ *Pantherophis guttatus* TL 180cm
(Red Rat Snake; Chicken Snake)

DESCRIPTION Orange-brown above, with black-edged, red, rhomboid-shaped vertebral blotches. Lower flanks often yellowish. Striking underside has black-and-white checkered pattern. MB 27–29 rows, 215–240 VENT, SUB 61–79 mostly divided and anal divided. **DISTRIBUTION** Introduced by accidental escapes and deliberate releases from illegal reptile collections. Most individuals found around Sydney, NSW, others around Melbourne and Brisbane. Native to south-eastern USA. **HABITAT AND HABITS** Terrestrial but

very adaptable, climbing trees while hunting. Lives in open forests, swamps and urban areas. Nocturnal, but basks by day. Feeds mainly on mammals, and birds and their eggs. Juveniles eat lizards and frogs. Lays 3–40 eggs. Large numbers of adults and juveniles found suggest that this species has become established in the wild. If seen, authorities should be notified. HARMLESS

Australian Slaty-grey Snake ■ *Stegonotus australis* TL 180cm

DESCRIPTION Black, dark grey or dark brown above, and lower flanks pale grey. White to cream below. MB 17 (or rarely 19) rows, 170–225 VENT, SUB 65–105 all divided and anal single. **DISTRIBUTION** Across northern Australia from Wadeye, NT, across Top End, NT, on to Cape York and south to Sarina, Qld. Also PNG and parts of Indonesia. **HABITAT AND HABITS** Terrestrial, living in forests, along waterways in savannah, swamps and floodplains. Commonly enters houses. Nocturnal. Feeds mainly on reptile eggs, frogs, mammals and fish. Has enlarged teeth for slicing open reptile eggs. Lays 7–16 eggs. Australian animals formerly thought to be *S. parvus* have been re-examined and assigned to this species. Large teeth inflict deep cuts that sting intensely and take significant time to form a blood clot, indicating that saliva may be toxic. HARMLESS

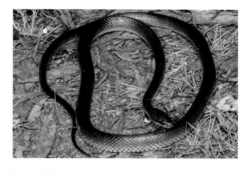

NATRICIDAE (WATER SNAKES)

The water snakes are a group of both solid-toothed and rear-fanged snakes that are mildly venomous. For many years they were thought to be a subfamily of the Colubridae. Much more diverse in Southeast Asia, only one of the approximately 238 species in the family occurs in northern Australia.

Keelback ■ *Tropidonophis mairii* TL 90cm
(Fresh-water Snake)

DESCRIPTION Grey to all shades of brown above, with darker flecks and spotting. White, yellow, orange to greenish with darker spotting below. MB 15 (or rarely 17) rows, 130–165 VENT, SUB 50–85 all divided and anal single. Distinguished from the Rough-scaled Snake (see p. 130) by having a loreal scale and corners of mouth upturned, forming a 'smile'. Many dorsal scales keeled, giving rise to its name. **DISTRIBUTION** Across northern and eastern Australia from Kimberley, WA, to Grafton, NSW. Also PNG and parts of Indonesia.

HABITAT AND HABITS Terrestrial, living in forests, and along waterways in savannah, swamps and floodplains. Commonly enters gardens. Cathemeral. Feeds mainly on frogs, lizards and fish. Known to eat small Cane Toads, although this takes a significant toll on the snakes' physiology to metabolize the toad toxins. Lays 3–18 eggs. Enlarged rear teeth can inflict deep cuts. HARMLESS

ELAPIDAE (FRONT-FANGED VENOMOUS SNAKES)

This family comprises venomous snakes that give birth to live young (sea snakes and some terrestrial elapids) or lay eggs (most terrestrial elapids). Many older texts place sea snakes and sea kraits in separate families. The Elapidae contains some of the world's best-known venomous snakes, such as the Black Mamba, Taipan, Tiger Snake, Eastern Coral Snake, Indian Cobra and Common Krait. Some have amazing toxicity – the Inland Taipan is the world's most toxic snake to mice. Bites on humans typically only occur when they try to pick up, provoke or kill a snake, or unwittingly stand on it when walking through long grass or dense ground litter. Elapids are Australia's most diverse group of snakes. A number of species and genera are poorly understood, and taxonomic changes are likely.

Common Death Adder ▪ *Acanthophis antarcticus* TL 75cm

DESCRIPTION Top of body reddish-brown to charcoal-grey, with lighter cross-bands. Lips white with black or dark grey markings. Belly colour similar to that of lighter bands, with darker flecking. Tail terminates in soft spine that is white to yellow or occasionally black. Stocky snake with raised scales above eyes. MB 21–23 rows, 110–135 VENT, SUB 35–60 single but posteriorly divided and anal single. **DISTRIBUTION** Southern WA, along southern coast to SA, and through most of NSW into north-east Qld. **HABITAT AND HABITS** Found in dry forests, rainforest margins, mallee, heaths, brigalow and grassland. Nocturnal, and seen by day while lying in ambush. Settles down in leaf litter and loose soil to ambush a prospective meal. Tail usually placed near head and wiggled to give appearance of a grub or worm, luring its lizard, frog, bird or mammal prey closer. Gives birth to 2–32 young. Unlike most Australian snakes, death adders often do not move away at the approach of a predator, relying on their superb camouflage. Death Adder or Polyvalent Antivenom is used to neutralize bites from this species. DANGEROUSLY VENOMOUS

Barkly Tableland Death Adder ■ *Acanthophis hawkei* TL 90cm

DESCRIPTION Top of body sandy-yellow to charcoal-grey, with lighter cross-bands. Edges of scales in cross-bands much brighter – when a snake flattens out indicating that it is threatened, the bright colours appear, startling the predator. Lips white with smudged darker markings. Belly colour similar to that of lighter bands, with darker flecking. Tail terminates in soft spine that is white to orange or occasionally black. MB 23 rows, 110–155 VENT, SUB 35–60 single but posteriorly divided and anal single. **DISTRIBUTION** From

south of Darwin, NT, to Longreach, Qld, via Gulf of Carpentaria. **HABITAT AND HABITS** Found on black-soil plains, and in grassland, floodplains and swamps. Predominantly nocturnal. Feeds on birds, lizards, frogs and mammals. Gives birth to 8–27 young. Death Adder or Polyvalent Antivenom is used to neutralize bites from this species. DANGEROUSLY VENOMOUS

North-western Death Adder ■ *Acanthophis lancasteri* TL 45cm

DESCRIPTION Top of body reddish-brown to orange, or greyish with lighter cross-bands. Lips cream with brown to grey markings. Belly cream to yellow. Tail terminates in soft spine that is white to black. MB 22–23 rows, 125–139 VENT, SUB 46–56 single but posteriorly divided and anal divided. **DISTRIBUTION** Kimberley region, WA; possibly extends into neighbouring areas of NT. **HABITAT AND HABITS** Lives in savannah woodland in close association with rocky outcrops. Nocturnal. Gives birth to 27 young. Species was recently revised with a description giving it the name *A. cryptamydros*, but it was found to have been previously named *A. lancasteri*. Death Adder or Polyvalent Antivenom is used to neutralize bites from this species. DANGEROUSLY VENOMOUS

Northern Death Adder ■ *Acanthophis praelongus* TL 45cm

DESCRIPTION Top of body yellow to reddish-brown to dark grey, with lighter cross-bands. Lips white with black or dark grey markings. Belly colour similar to that of lighter bands, with darker flecking. Tail terminates in soft spine that is white to orange or occasionally black. Stocky snake with the highest raised supraoculars of all Australian death adders. MB 19–21 rows, 110–135 VENT, SUB 35–60 single but posteriorly divided and anal single. **DISTRIBUTION** Restricted to north-east Qld, from Torres Strait Islands to Whitsunday region. **HABITAT AND HABITS** Lives in rainforests, grassland and woodland. Predominantly nocturnal. Feeds on lizards, frogs and small mammals. Gives birth to 6–17 young. Death Adder or Polyvalent Antivenom is used to neutralize bites from this species. DANGEROUSLY VENOMOUS

Desert Death Adder ■ *Acanthophis pyrrhus* TL 75cm

DESCRIPTION Top of body orange to reddish-brown, with lighter cross-bands. Lips peppered orange-brown with lower margins white. Belly white to cream. Tail terminates in soft spine that is white, yellow or occasionally black. MB 19–21 rows, 120–162 VENT, SUB 45–67 single but last few divided and anal single. **DISTRIBUTION** Western and

central Australia. **HABITAT AND HABITS** Lives in sand-ridge and rocky desert regions and associated adjoining habitats, with a strong preference for spinifex. Predominantly nocturnal. Feeds mainly on skinks and occasionally on mammals. Gives birth to 9–14 young. Death Adder or Polyvalent Antivenom is used to neutralize bites from this species. DANGEROUSLY VENOMOUS

Papuan Death Adder ■ *Acanthophis rugosus* TL 75cm
(Woodland Death Adder)

DESCRIPTION Top of body yellow to charcoal-grey, with lighter cross-bands. Edges of scales in cross-bands are much brighter, and when a snake flattens out indicating that it is scared or threatened, bright colours appear, startling the predator. Lips white with black or dark grey markings. Belly colour similar to that of lighter bands with darker flecking. Tail terminates in soft spine that is white to orange or occasionally black. Raised scales above eyes and keeled head shields. MB 23 rows, 115–165 VENT, SUB 53 single but posteriorly divided and anal single.

DISTRIBUTION Across northern Australia from Kimberley region into western Qld. Also PNG.

HABITAT AND HABITS Found in forests and grassland with rocky soils. Feeds mainly on skinks and occasionally mammals. Gives birth to 6–24 young. A species complex; Selwyn Range animals are distinct but further research is needed to determine their status. Death Adder or Polyvalent Antivenom is used to neutralize bites from this species. DANGEROUSLY VENOMOUS

Selwyn Range population

Pilbara Death Adder ■ *Acanthophis wellsei* TL 55cm

DESCRIPTION Top of body pale brown to orange, with lighter cross-bands; another colour phase is red with black bands and black head. Belly colour similar to that of lighter bands with darker flecking. Tail terminates in soft spine that is white or occasionally black. MB 17–21 rows, 119–143 VENT, SUB 41–64 single but posteriorly divided and anal single.

Cape Range population

DISTRIBUTION Pilbara region and Cape Range area, **WA. HABITAT AND HABITS** Rocky desert and gorges and associated adjoining habitats, with strong preference for spinifex. Feeds mainly on skinks and occasionally mammals. Gives birth to 9–20 young. Possibly a species complex; Cape Range animals are distinctive but require further investigation. Death Adder or Polyvalent Antivenom is used to neutralize bites from this species. DANGEROUSLY VENOMOUS

Short-nosed Sea Snake ■ *Aipysurus apraefrontalis* TL 110cm

DESCRIPTION Top of body beige to dark brown, with dark brown to purplish-grey cross-bands that taper on flanks. Underside pale beige to cream. MB 17 rows, 140–160 VENT, SUB 18–35 all single and anal divided. Differs from the Leaf-scaled Sea Snake (see p. 44) by having fewer midbody scales (17 vs 19–21). **DISTRIBUTION** Waters in WA from Exmouth to Ashmore Reef. **HABITAT AND HABITS** Lives in waters over both coral reefs and rocky areas. Recorded being found under pieces of broken coral at low tide on Ashmore Reef. Nocturnal. Known to eat gobies and eels. Thought to be livebearer. Sea Snake Antivenom is expected to neutralize bites from this species. VENOMOUS

Dubois' Sea Snake ■ *Aipysurus duboisii* TL 120cm

DESCRIPTION Top of body cream to white, with grey to black bands. Edges of midbody scales white, forming reticulated pattern on body. Belly colour similar to dorsal surface. MB 17 rows, 154–181 VENT, SUB 25–30 all divided and anal divided. **DISTRIBUTION** Waters around Australia, north of Ballina, NSW, to Exmouth, WA. Also western Pacific to New Caledonia. **HABITAT AND HABITS** Lives in waters over coral reefs, mudflats and seagrass beds. Nocturnal. Feeds on fish. Gives birth to 2–7 live young. Sea Snake Antivenom is expected to neutralize bites from this species. DANGEROUSLY VENOMOUS

Stagger-banded Sea Snake ■ *Aipysurus eydouxii* TL 91cm
(Marbled Sea Snake)

DESCRIPTION Top of body pale cream to light grey, with 29–45 dark bands. Belly black to cream with dark midline. MB 17 rows, 127–149 VENT, SUB 21–34 all single and anal divided. Differs from the Mosaic Sea Snake (see p. 46) by having more midbody bands (29–45 vs 15–22).

DISTRIBUTION Extralimital in Australian waters. Known from single individual listed from Port Moresby, PNG, on Sahul Shelf, and therefore presumed to occur in Torres Strait, Qld. Occurs extensively throughout Southeast Asia. **HABITAT AND HABITS** Found in turbid waters up to 60m deep. Nocturnal. Eats fish eggs; this diet specialization has led to an atrophied venom apparatus. Gives birth to 4 young. HARMFUL

Leaf-scaled Sea Snake ■ *Aipysurus foliosquama* TL 50cm

DESCRIPTION Top of body brown to dark brown, sometimes with obscure darker cross-bands that taper on flanks. Underside grey-brown sparsely flecked with white. MB 19–21 rows, 139–153 VENT, SUB 20–29 all single and anal divided. Differs from the Short-nosed Sea Snake (see p. 43) by having more midbody scales (19–21 vs 17). **DISTRIBUTION**

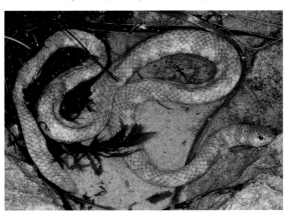

Waters in WA from Hibernia and Ashmore Reefs. Also localized population at Shark Bay. **HABITAT AND HABITS** Lives in waters over coral and rocky reefs, and in Shark Bay over seagrass beds. Nocturnal. Eats small fish. Thought to be livebearer. Sea Snake Antivenom is expected to neutralize bites from this species. VENOMOUS

Dusky Sea Snake ■ *Aipysurus fuscus* TL 94cm

DESCRIPTION Top of body dark chocolate-brown. Belly colour similar to upper surface. MB 17 rows, 156–172 VENT, SUB 24–37 all single and anal divided. Differs from the Olive Sea Snake (see below) by having fewer midbody scales (17 vs 21–25).

DISTRIBUTION Waters in WA from Ashmore and Scott Reefs in Timor Sea. **HABITAT AND HABITS** Lives in waters over coral reefs and reef flats. Predominantly nocturnal. Eats small fish and gobies. Gives birth to live young. Sea Snake Antivenom is expected to neutralize bites from this species. DANGEROUSLY VENOMOUS

Olive Sea Snake ■ *Aipysurus laevis* TL 170cm
(Golden Sea Snake)

DESCRIPTION Top of body grey-brown to white, with or without darker and lighter blotches and specks. Some individuals white with golden, apricot or dark brown head. Belly colour similar to upper surface. MB 21–25 rows, 142–156 VENT, SUB 22–30 all single and anal divided. Differs from the Dusky Sea Snake (see above) by having more midbody scales (21–25 vs 17). **DISTRIBUTION** Waters around Australia, north

of Sydney, NSW, to Exmouth, WA. Also New Caledonia and southern PNG. **HABITAT AND HABITS** Lives in waters over coral reefs and rocky areas. Diurnal. Eats wide range of small fish, fish eggs, and occasionally prawns and crabs. Gives birth to 1–5 young. Sea Snake Antivenom is expected to neutralize bites from this species. DANGEROUSLY VENOMOUS

Mosaic Sea Snake ■ *Aipysurus mosaicus* TL 110cm

DESCRIPTION Top of body pale cream to light grey, with 15–22 dark bands. Belly black

to cream with dark midline. MB 17 rows, 140–157 VENT, SUB 25–38 all single and anal divided. Differs from the Stagger-banded Sea Snake (see p. 44) by having fewer midbody bands (15–22 vs 29–45). **DISTRIBUTION** From Exmouth, WA, across northern Australia, to Brisbane, Qld. Waifs occur to Sydney, NSW. Also recorded off southern PNG. **HABITAT AND HABITS** Found in turbid waters up to 50m deep. Nocturnal. Eats fish eggs, with this diet specialization leading to an atrophied venom apparatus. Gives birth to 5 young. HARMFUL

Shark Bay Sea Snake ■ *Aipysurus pooleorum* TL 114cm

DESCRIPTION Top of body grey to black, with pale grey to cream bands. Belly colour similar to upper surface. Juveniles much brighter with cleaner markings than adults. MB 20–23 rows, 146–159 VENT, SUB 25–33 all single and anal divided. **DISTRIBUTION** Waters around Shark Bay, occasionally straying to Perth, WA. **HABITAT AND HABITS** Lives in waters over coral and rocky reefs, and over seagrass beds in Shark Bay. Cathemeral. Eats small fish. Thought to be livebearer. Sea Snake Antivenom is expected to neutralize bites from this species. DANGEROUSLY VENOMOUS

Mjoberg's Sea Snake ■ *Aipysurus tenuis* TL 130cm

DESCRIPTION Top of body cream to pale grey or brown, sometimes with dark brown speckles forming longitudinal stripes and specks. Golden, apricot or dark brown head. Belly colour similar to upper surface. MB 19 rows, 185–194 VENT, SUB 35–40 all single and anal divided.

DISTRIBUTION Waters from Broome to Dampier Archipelago, WA. HABITAT AND HABITS Lives in waters over coral and rocky reefs. Cathemeral. Known to eat small fish. Thought to be livebearer. Sea Snake Antivenom is expected to neutralize bites from this species. DANGEROUSLY VENOMOUS

North-eastern Plain-nosed Burrowing Snake
■ *Antaioserpens albiceps* TL 42cm
(Warro)

DESCRIPTION Top of body grey-brown to reddish-orange, with lighter yellowish centres on each of midbody scales. Head and lips whitish and strongly peppered with dark grey markings. Back of head has creamy-yellow bar, with broad black band on nape. Belly creamish-white. Compared to adults, juveniles have much broader white stripe on nape that covers much more of head. MB 15 rows, 135–165 VENT, SUB 15–25 and anal divided. DISTRIBUTION Restricted to north-east Qld, from Clermont to northern Cape York Peninsula. HABITAT AND HABITS Lives in grassland and woodland. Predominantly nocturnal. Feeds on lizards. Lays 3 eggs. HARMFUL

Robust Burrowing Snake ■ *Antaioserpens warro* TL 44cm

DESCRIPTION Top of body grey-brown to reddish-orange, with random dark brown groupings of scales. Head and lips whitish and strongly peppered with dark grey markings. Back of head has faded orange bar, with broad black band on nape. Belly creamish-white. Juveniles probably brighter in colour than adults. MB 15 rows, 139–150 VENT, SUB 15–17 and anal divided. Recently split from the North-eastern Plain-nosed Burrowing

Snake (see p. 47); differs by having dark speckling on top of body. **DISTRIBUTION** Restricted to central and eastern Qld, from Mitchell to Morven and Charleville. Also recorded from Blackdown Tablelands and around Port Curtis. **HABITAT AND HABITS** Lives in woodland, mulga and grassland. Predominantly nocturnal. Feeds on lizards. Thought to lay eggs. HARMFUL

Pygmy Copperhead ■ *Austrelaps labialis* TL 75cm

DESCRIPTION Top of body grey or dark brown, the lips boldly marked with black or dark grey on white. Belly cream to yellow, with occasional orange flecks. Robust snake with scales that are matt in finish. MB 15 (rarely 17) rows, 133–155 VENT, SUB 35–58 and anal single. **DISTRIBUTION** South-east SA, in Adelaide Hills and on Kangaroo Island. **HABITAT**

AND HABITS Usually found under cover or among grass tussocks in swamps, grassland and moist forests. Diurnal, but crepuscular in hot weather. Feeds mainly on frogs and lizards, but will also eat other snakes and occasionally mammals. Gives birth to 3–32 young. Tiger Snake or Polyvalent Antivenom is used to neutralize bites from this species. DANGEROUSLY VENOMOUS

Highland Copperhead ■ *Austrelaps ramsayi* 130cm
(Alpine Copperhead)

DESCRIPTION Top of body light brown to charcoal-grey, and thin vertebral stripe in some individuals. Lips boldly marked with black or dark grey markings on white background; this distinguishes it from the Lowland Copperhead (see p. 50). Belly cream to yellow with occasional orange flecks. Copperheads are robust snakes with scales that are matt in finish. MB 15 (rarely 17) rows, 150–171 VENT, SUB 35–58 and anal single. **DISTRIBUTION** South-eastern Australia from eastern Vic, up eastern seaboard, to Qld border. Distribution becomes patchier in north of range. **HABITAT AND HABITS** Lives in forests, grassland,

heaths, swamps and riverine systems. Also readily exploits human disturbed environments such as rural areas and suburban gardens. Diurnal, and occasionally nocturnal in warm weather. Feeds on lizards, frogs and small mammals. Gives birth to 5–31 young. Tiger Snake or Polyvalent Antivenom is used to neutralize bites from this species. DANGEROUSLY VENOMOUS

Lowland Copperhead ■ *Austrelaps superbus* 150cm
(Superb Snake)

DESCRIPTION Top of body light orange-brown to charcoal-grey. Lateral stripe invariably present, which is usually yellow to copper coloured, and often extends up on to nape, hence the name copperhead. Some individuals also have spots dotted along body. Lips have dark markings on white background. These do not have a sharp delineation like those of the Pygmy and Highland Copperheads (see pp. 48 and 49). Belly cream to yellow, with occasional orange flecks. Copperheads are robust snakes with scales that are matt in finish. MB 15 (rarely 17) rows, 140–165 VENT, SUB 35–58 and anal single. **DISTRIBUTION** From SA, across Vic and Tas, including some offshore islands, and into southern NSW.

HABITAT AND HABITS Lives in forests, grassland, heaths, swamps and riverine systems. Also uses urban environments. Diurnal, but occasionally nocturnal in warm weather. Feeds on lizards, frogs and small mammals. Gives birth to 2–32 young. Tiger Snake or Polyvalent Antivenom is used to neutralize bites from this species. DANGEROUSLY VENOMOUS

North-western Shovel-nosed Snake
■ *Brachyurophis approximans* TL 37cm

DESCRIPTION Top of body dark brown to charcoal with thin, one scale-wide cream to

pale grey irregular cross-bands. Belly creamish-white. MB 17 rows, 151–181 VENT, SUB 19–27 and anal divided. **DISTRIBUTION** WA, from Pilbara region, south to Yalgoo in goldfields region. **HABITAT AND HABITS** Lives on heavy clay to rocky soils. Shelters beneath timber in Mulga woodland and heath associations. Nocturnal. Feeds only on reptile eggs. Lays 2–4 eggs. HARMFUL

Australian Coral Snake ■ *Brachyurophis australis* TL 50cm

DESCRIPTION Top of body reddish-brown, pink to orange, with narrow, darkened cross-bands. These have cream to yellow centres in midbody scales. Conspicuous dark nape-blotch or band that is wider than following bands. Belly creamish-white. Snout blunt and upturned, an adaptation to its fossorial habits. MB 17 rows, 140–170 VENT, SUB 15–31 and anal divided. **DISTRIBUTION** From Mingela, Qld, south through Qld mainly west of GDR, through NSW, north-

west Vic and across to Port Pirie, SA. **HABITAT AND HABITS** Lives on heavy clay to rocky or sandy soils in mulga, brigalow, open woodland and mallee associations. Found under rocks and fallen timber, and beneath leaf litter. Nocturnal. Feeds on lizards and reptile eggs. Lays 4–6 eggs. HARMFUL

Einasleigh Shovel-nosed Snake ■ *Brachyurophis campbelli* TL 39cm

DESCRIPTION Top of body pale grey to yellowish-pink to orange, with dark brown to charcoal, broad, irregular cross-bands. Conspicuous dark nape-blotch or mark that is broader in depth than following bands. Belly creamish-white. Snout blunt and upturned, an adaptation to its fossorial habits. MB 15–17 rows, 140–190 VENT, SUB 14–30 and anal divided. **DISTRIBUTION** Qld from western Cape York Peninsula, south to Townsville and west to Camooweal.

HABITAT AND HABITS Shelters beneath timber and rocks in open tropical woodland and on rocky hillsides. Nocturnal. Feeds only on reptile eggs. Lays 6 eggs. Population with 15 midbody scale rows, found in eastern part of range, has name *woodjonesii* available. HARMFUL

Western Narrow-banded Shovel-nosed Snake
■ *Brachyurophis fasciolatus fasciolatus* TL 37cm

DESCRIPTION Top of body white to cream with pale pink to orange flecks, and regular, thick, dark brown to charcoal, ragged-edged cross-bands. Conspicuous dark nape-blotch or band that is wider than following bands. Belly creamish-white. Snout blunt and upturned, an adaptation to its fossorial habits. MB 17 rows, 140–175 VENT, SUB 15–30 and anal

divided. The broader dark bands, along with distribution, separate it from the Eastern Narrow-banded Shovel-nosed Snake (see below). **DISTRIBUTION** WA, from Perth east to Laverton in goldfields region, and north to Shark Bay. **HABITAT AND HABITS** Lives on sandy soils. Shelters beneath cover and leaf litter in mulga and mallee woodland and heaths, and inside stick-ant nests. Nocturnal. Feeds on reptile eggs. Lays 2–5 eggs. HARMFUL

Eastern Narrow-banded Shovel-nosed Snake
■ *Brachyurophis fasciolatus fasciata* TL 41cm

DESCRIPTION Top of body white to cream with pale pink to orange flecks, and regular, thin, dark brown to charcoal, ragged-edged cross-bands. Conspicuous dark nape-blotch or band that is wider than following bands. Belly creamish-white. Snout blunt and upturned, an adaptation to its fossorial habits. MB 17 rows, 140–171 VENT, SUB 19–27 and anal

divided. Thin dark bands, along with distribution, separate it from the Western Narrow-banded Shovel-nosed Snake (see above). **DISTRIBUTION** Eastern WA, through central Australia, to western NSW and Qld. Isolated population in Little and Great Sandy Deserts, WA, may represent an undescribed taxon. **HABITAT AND HABITS** Lives on sandy soils. Shelters beneath cover and leaf litter in mulga woodland, heaths, sand-ridge deserts and mallee. Nocturnal. Feeds only on reptile eggs. Lays 4–7 eggs. HARMFUL

Unbanded Shovel-nosed Snake ■ *Brachyurophis incinctus* TL 32cm

DESCRIPTION Top of body reddish-brown, pink to orange, without characteristic bands of other genus members. Conspicuous dark nape-blotch or band, usually with additional dark band over eyes. Belly creamish-white. Snout blunt and upturned, an adaptation to its fossorial habits. MB 17 rows, 140–165 VENT, SUB 18–31 and anal divided.

DISTRIBUTION From Quilpie, Qld, north to Mt Isa, and south-west across Barkly Tableland to Alice Springs, NT. **HABITAT AND HABITS** Lives on heavy clay and sandy soils in mulga, open woodland and sand-ridge deserts. Found under rocks and fallen timber, and beneath leaf litter. Nocturnal. Feeds only on reptile eggs. Lays 3–5 eggs. HARMFUL

Arnhem Shovel-nosed Snake ■ *Brachyurophis morrisi* TL 30cm

DESCRIPTION Top of body reddish-brown to brown, without characteristic bands of other genus members. Conspicuous dark nape-blotch or band, usually with additional dark band over eyes. Usually dark orange between bands. Belly creamish-white. Snout blunt and upturned, an adaptation to its fossorial habits. MB 15 rows, 135–145 VENT, SUB 15–25 and anal divided.

DISTRIBUTION Northern NT, from Elcho Island and Cobourg Peninsula, to Nabarlek. **HABITAT AND HABITS** Lives on heavy clay and sandy soils in tropical woodland under rocks and fallen timber, and beneath leaf litter. Nocturnal. Probably feeds on reptile eggs. Thought to lay eggs. HARMFUL

Northern Shovel-nosed Snake ■ *Brachyurophis roperi* TL 37cm

DESCRIPTION Top of body dark brown to charcoal coloured. Thin, irregular cross-bands, which are variable at 1–5 scales wide. Bands are cream, yellow-orange or pinkish. Belly creamish-white. Snout blunt and upturned, an adaptation to burying itself in the substrate. MB 15–17 rows, 150–180 VENT, SUB 15–24 and anal divided. **DISTRIBUTION** From Kimberley, WA, across to Booroloola, NT and to the Qld border region near Camooweal. **HABITAT AND HABITS** Lives on heavy clay to rocky soils. Shelters beneath timber and rocks in open tropical woodland and on rocky hillsides. Nocturnal. Feeds only on reptile eggs. Lays up to 5 eggs. HARMFUL

Southern Shovel-nosed Snake ■ *Brachyurophis semifasciatus* TL 37cm

DESCRIPTION Top of body reddish-brown to charcoal coloured, with regular cross-bands that are variable at 2–3 scales wide. Bands are cream, yellow-orange or red. Belly creamish-white. Snout blunt and upturned, an adaptation to its fossorial habits. MB 17 rows, 147–188 VENT, SUB 15–26 and anal divided. **DISTRIBUTION** Southern WA, between Cockburn and Wiluna, across to south-west NT, and south to Eyre Peninsula, SA. **HABITAT AND HABITS** Lives on heavy clay to sandy soils. Shelters beneath timber and rocks in open woodland, mallee, sand-ridge deserts and heaths. Nocturnal. Feeds only on reptile eggs. Lays 1–7 eggs. HARMFUL

Northern Crowned Snake ■ *Cacophis churchilli* TL 50cm

DESCRIPTION Top of body charcoal-grey to blackish. Some individuals have yellow longitudinal stripes on first third of body. Top of head dark, while sides and rear are stippled with grey, brown and white. Conspicuous yellow or white band across nape. Belly grey. MB

15 rows, 160–175 VENT, SUB 20–30 and anal divided. Differs from the White-crowned Snake (see below) by having lower ventral scale count.
DISTRIBUTION Restricted to north-east Qld, from Mossman south to Paluma. Also recorded from the Whitsundays, south to Sarina. **HABITAT AND HABITS** Lives in rainforests and woodland. Nocturnal. Feeds on lizards. Lays 7–9 eggs. HARMFUL

White-crowned Snake ■ *Cacophis harriettae* TL 55cm

DESCRIPTION Top of body charcoal-grey to blackish. Top of head dark, while sides and rear are stippled with grey, brown and white. Conspicuous wide pale yellow or white band across nape. Belly grey. MB 15 rows, 170–200 VENT, SUB 25–45 and anal divided. Differs from the Northern Crowned Snake (see above) by having higher ventral scale count.
DISTRIBUTION Qld, from Kirrima south along coast to just south of Grafton, NSW. Also extends west as far as Glenmorgan, Qld. **HABITAT AND HABITS** Lives in rainforests, woodland and gardens. Nocturnal. Feeds on lizards. Lays 2–10 eggs. HARMFUL

Southern Dwarf Crowned Snake ■ *Cacophis krefftii* TL 32cm

DESCRIPTION Top of body charcoal-grey to blackish. Top of head dark, while sides and rear are stippled with grey, brown and white. Conspicuous yellow, orange or white thin band across nape. Belly yellow, distinguishing it from other crowned snakes. MB 15 rows, 140–160 VENT, SUB 20–30 and anal divided. **DISTRIBUTION** Qld, from Whitsundays south to Gosford, NSW. **HABITAT AND HABITS** Lives in rainforests, woodland and gardens. Nocturnal. Feeds on lizards. Lays 2–5 eggs. HARMFUL

Golden-crowned Snake ■ *Cacophis squamulosus* TL 75cm

DESCRIPTION Top of body tan to brownish-black. Top of head dark, while sides and rear are stippled with grey, brown and white. Conspicuous orange to yellowish marking that often extends down from around head along forebody. Belly orange with black flecks, distinguishing it from other crowned snakes. MB 15 rows, 165–185 VENT, SUB 25–40 and anal divided.

DISTRIBUTION Coastal Qld, from Eungella south to Wollongong, NSW. **HABITAT AND HABITS** Lives in rainforests, woodland and gardens. Often found under cover such as rotting logs and rocks, or when brought into homes by cats. Nocturnal. Feeds on lizards. Lays 2–15 eggs. HARMFUL

Carpentaria Snake ▪ *Cryptophis boschmai* TL 45cm

DESCRIPTION Top of body blackish-brown to tan, usually lighter on flanks than on midline. This can form a longitudinal stripe running along lower flanks. Belly white. MB 15 rows, 145–190 VENT, SUB 20–35 and anal single. Similar to the Eastern Small-eyed Snake (see p. 57), but differs by nasal scale not being in contact with preocular scale. **DISTRIBUTION** Qld, from western Cape York Peninsula to outskirts of Brisbane. Also New Guinea. **HABITAT AND HABITS** Lives in woodland, brigalow and tropical savannah. Nocturnal. Feeds on lizards. Gives birth to 5–11 young. VENOMOUS

Pink Snake ▪ *Cryptophis incredibilis* TL 45cm

DESCRIPTION Top of body bright pink, as the name suggests. Belly whitish-cream. MB 15 rows, 180–185 VENT, SUB 50–65 and anal single. Similar to the Black-striped Snake (see p. 58), but differs due to lack of dark vertebral stripe. **DISTRIBUTION** Only on Prince of Wales Island in Torres Strait. **HABITAT AND HABITS** Most individuals have been found beneath flotsam on a shoreline that adjoins open woodland on sandy soil, or while actively hunting at night. Nocturnal. Presumably feeds on lizards and gives birth to live young. Possibly a colour variant of the Black-striped Snake. HARMFUL

Eastern Small-eyed Snake ■ *Cryptophis nigrescens* TL 80cm

DESCRIPTION Top of body grey to jet-black, with black head. Belly pink to orange-red, sometimes with grey flecks. MB 15 rows, 165–210 VENT, SUB 30–47 and anal single. Similar to the Carpentaria Snake (see p. 57), differing by nasal scale being in contact with preocular scale. Also confused with the Red-bellied Black Snake (see p. 114), differing by lacking clear black bands across belly. Older individuals can develop a condition called macrocephaly, where the head enlarges past its normal size. **DISTRIBUTION** From

Mossman, Qld, along GDR, to Melbourne, Vic. **HABITAT AND HABITS** Lives in forests, woodland and rocky heaths. Nocturnal. Diet consist of lizards, frogs and occasionally other snakes. Gives birth to 2–8 young. Tiger Snake or Polyvalent Antivenom is used to neutralize bites from this species. DANGEROUSLY VENOMOUS

Black-striped Snake ■ *Cryptophis nigrostriatus* TL 50cm

DESCRIPTION Top of body charcoal-grey to blackish, with a pair of broad pink stripes on flanks. Width of stripes is variable. Top of head dark. Belly whitish-cream. MB 15 rows,

160–190 VENT, SUB 45–75 and anal single. Similar to the Pink Snake (see p. 57), but differs by having dark vertebral stripe. **DISTRIBUTION** Eastern Qld, from Awoonga Dam, north to tip of Cape York Peninsula. Also PNG. **HABITAT AND HABITS** Lives in rainforests and woodland. Nocturnal. Feeds on lizards. Gives birth to 4–9 young. HARMFUL

Northern Small-eyed Snake ▪ *Cryptophis pallidiceps* TL 50cm
(Secretive Snake)

DESCRIPTION Top of body blackish-brown. Longitudinal pinkish-orange to yellow stripe runs along lower flanks. Belly white. MB 15 rows, 160–180 VENT, SUB 35–55 and anal single. **DISTRIBUTION** WA, from Kimberley region across northern section of NT, to Groote Eylandt. **HABITAT AND HABITS** Lives in rocky woodland and tropical savannah. Nocturnal. Feeds on lizards. Gives birth to 2–5 young. Bites can cause significant swelling and pain, and have resulted in marked symptoms. VENOMOUS

Narrow-headed Whip Snake ▪ *Demansia angusticeps* TL 90cm

DESCRIPTION Top of body brownish-olive to grey. Anterior edge of each midbody scale can be marked with black. Eye surrounded by dark brown, comma-shaped mark. Pale-edged brown line runs between nostrils, bisecting rostral scale. Lower flanks can have a yellow flush. Belly white, cream or yellow, often with grey flecks anteriorly. MB 15 rows, 180–200 VENT, SUB 70–100 and anal divided. **DISTRIBUTION** WA, from Kimberley region, across far north-western NT. **HABITAT AND HABITS** Lives in rocky woodland and tropical savannah. Diurnal. Feeds on lizards. Probably lays eggs. Bites can cause significant swelling and pain. VENOMOUS

Black-necked Whip Snake ■ *Demansia calodera* TL 70cm

DESCRIPTION Top of body grey to reddish-brown. Usually has conspicuous black band edged with cream or white on nape that may fade to grey in old individuals. Anterior edge of each midbody scale can be marked with black. Eye surrounded by dark brown, comma-

shaped mark. Pale-edged brown line runs between nostrils, bisecting rostral scale. Lower flanks can have a yellow flush. Belly white, cream or pale yellow. Some individuals reported to have dark posterior edges to ventral scales. MB 15 rows, 170–195 VENT, SUB 65–90 and anal divided. **DISTRIBUTION** WA, from Kalbarri to North West Cape. Also in Gibson and Little Sandy Deserts. **HABITAT AND HABITS** Lives in rocky woodland and tropical savannah. Diurnal. Feeds on lizards. Probably lays eggs. Bites can cause significant swelling and pain. HARMFUL

Carpentarian Whip Snake ■ *Demansia flagellatio* TL 71cm
(Ornate Whip Snake)

DESCRIPTION Top of body reddish-brown to bluish-grey. Head and nape dark grey to charcoal, with pale sides to head and 2 broad, pale orange to yellow bands. Eye surrounded by dark grey, comma-shaped mark. Dark line runs between nostrils, bisecting rostral

scale. Rear of body gradually becomes bright yellow. Ventral colour white, cream or yellow. MB 15 rows, 195–215 VENT, SUB 100–115 and anal divided. **DISTRIBUTION** North-western region of Qld along Selwyn Range. **HABITAT AND HABITS** Lives in tropical savannah and spinifex-dominated rocky woodland. Diurnal. Feeds on lizards. Probably lays eggs. Bites can cause significant swelling and pain, but may result in more serious symptoms. HARMFUL

Olive Whip Snake ■ *Demansia olivacea* TL 80cm
(Marbled-headed Whip Snake)

DESCRIPTION Top of body brownish-olive to grey. Anterior edge of each midbody scale can be marked with black. Head finely stippled with grey or brown. Dark-coloured streak from eye back along body towards corner of mouth. Belly white, cream or yellow, often with grey flecks anteriorly. MB 15 rows, 160–210 VENT, SUB 65–110 and anal divided. **DISTRIBUTION** WA, from Kimberley region, across Top End region of NT, to edge of Gulf of Carpentaria. **HABITAT AND HABITS** Lives in rocky woodland and tropical savannah. Diurnal. Feeds on lizards. Lays 3–5 eggs. Bites can cause significant swelling and pain, but may result in more serious symptoms. VENOMOUS

Greater Black Whip Snake ■ *Demansia papuensis* TL 160cm

DESCRIPTION Top of body grey to black, but some individuals are tan to reddish-brown. Belly grey. MB 15 rows, 198–225 VENT, SUB 75–110 and anal divided. Similar to the Lesser Black Whip Snake (see p. 67), but differs by having more than 197 ventral scales. **DISTRIBUTION** From Mackay, up east coast and across north coast of Australia. Type specimen probably originated in Qld, and incorrectly thought to be from southern PNG. **HABITAT AND HABITS** Found in forests, open woodland and grassland, beneath logs, rocks and man-made debris. Diurnal, but can be nocturnal in hot weather. Feeds on lizards, frogs and occasionally other snakes. Lays 5–13 eggs. Polyvalent Antivenom is used to neutralize bites from this species. DANGEROUSLY VENOMOUS

Yellow-faced Whip Snake ■ *Demansia psammophis* TL 80cm

DESCRIPTION Top of body variable, from olive, grey, brown and greenish, to reddish-orange and grey. Head can be yellow or same colour as body. Some populations have black-edged midbody scales. North-eastern individuals often have a pair of reddish stripes on the first third, blending into grey towards rear of body. Belly white to greenish-yellow, and usually bright yellow under tail. Characteristic black, comma-shaped mark over eye,

and pale-edged brown line that runs between nostrils, bisecting rostral scale. MB 15 rows, 165–230 VENT, SUB 60–105 and anal divided. **DISTRIBUTION** From south-eastern SA, through north-west Vic, to NSW coast, and up eastern half of Qld to Mossman. **HABITAT AND HABITS** Lives in rocky woodland and tropical savannah. Diurnal. Feeds on lizards. Lays 2–17 eggs. Huge communal laying sites are known. Possibly more than one species is involved. Bites can cause significant swelling and pain, and have resulted in marked symptoms. VENOMOUS

Sombre Whip Snake ■ *Demansia quaesitor* TL 75cm

DESCRIPTION Brown to grey with black band across nape. Eastern population rusty-red to grey, with or without reddish head, but lacks black nape-band. Dark streak leading from eye towards rear of mouth. Ventral colouration white. MB 15 rows, 180–200 VENT, SUB 60–100 and anal divided. **DISTRIBUTION** WA, from eastern Kimberley region across NT, through south of Gulf of Carpentaria, and along Selwyn Range, Qld to Dajarra. **HABITAT AND HABITS** Lives in rocky woodland and tropical savannah. Diurnal. Feeds on lizards. Lays 6 eggs. Possibly more than one species is involved. Bites can cause significant swelling and pain, and have resulted in marked symptoms. HARMFUL

Western population

Eastern population

Reticulated Whip Snake ■ *Demansia reticulata reticulata* TL 100cm

DESCRIPTION First half of body greenish-grey to yellow. Remainder of body and tail usually brown or grey. Posterior edges of each midbody scale usually edged with black. Eye has white diagonal streak bordered posteriorly with black. Belly white to yellow. MB 15 rows, 165–217 VENT, SUB 70–102 and anal single. **DISTRIBUTION** WA, from Shark Bay south through western WA, to Myalup and inland to Kellerberrin. **HABITAT AND HABITS** Lives in heaths, mallee, mulga and open woodland. Diurnal. Feeds on lizards. Lays up to 6 eggs. Formerly thought to be a subspecies of the Yellow-faced Whip Snake (see opposite). Bites can cause significant swelling and pain, and have resulted in marked symptoms. VENOMOUS

Centralian Whip Snake ■ *Demansia reticulata cupreiceps* TL 75cm

DESCRIPTION Head yellowish-tan. First half of body greenish-grey. Remainder of body and tail the same yellowish-tan as head. Eye has white diagonal streak bordered posteriorly with black. Belly white to yellow. MB 15 rows, 175–200 VENT, SUB 60–105 and anal single. **DISTRIBUTION** Continuous from Eastern Goldfields, WA, through SA

to Renmark, and through far western NSW and south-west Qld. Isolated population occurs in Kimberley region. **HABITAT AND HABITS** Lives in sand-ridge deserts, rocky gorges, mallee, mulga and woodland. Diurnal. Feeds on lizards. Lays 5–8 eggs. Formerly thought to be a subspecies of the Yellow-faced Whip Snake (see p. 62). Bites can cause significant swelling and pain, and have resulted in marked symptoms. VENOMOUS

Crack-dwelling Whip Snake ■ *Demansia rimicola* TL 90cm
(Blacksoil Whip Snake)

DESCRIPTION Top of body brownish-olive to brown. Lower flanks can have reddish-brown stripes along lower flanks. Head and neck usually darker brown than body, particularly in young individuals. Usually 2 white bands both in front and behind eyes, as well as 2 bands on nape. Head finely stippled with grey or black. Dark-coloured streak from eye back along body towards corner of mouth. Belly orange to red, with 2 longitudinal dark brown spots forming parallel rows. MB 15 rows, 175–205 VENT, SUB 65–110 and anal

divided. **DISTRIBUTION** WA, from Kimberley region, across central and eastern NT, through into north-east SA. Also through western Qld to Roma, and north-west to Gregory. **HABITAT AND HABITS** Lives in Mitchell grassland, black-soil plains and tropical savannah. Diurnal. Feeds on lizards. Lays up to 8 eggs. Bites can cause significant swelling and pain, but may result in more serious symptoms. HARMFUL

Rufous Whip Snake ▪ *Demansia rufescens* TL 68cm
(Red Whip Snake)

DESCRIPTION Top of body reddish-brown. Head and neck grey. Eye has dark brown, comma-shaped mark around it. Belly white or cream. MB 15 rows, 177–200 VENT, SUB 65–85 and anal divided.

DISTRIBUTION WA, from Port Hedland, through Pilbara, to North West Cape. Also on Barrow and Dolphin Islands. **HABITAT AND HABITS** Lives in rocky woodland and gorges dominated by spinifex on rocky soils. Diurnal. Feeds on lizards. Probably lays eggs. Bites may cause significant swelling and pain, but may result in more serious symptoms. VENOMOUS

Shine's Whip Snake ▪ *Demansia shinei* TL 84cm

DESCRIPTION Top of body brown to bluish-grey. Head and nape grey with pale sides to head and 2 broad, pale orange to yellow bands. Broad yellow band behind eye over temporal region. Band thinner in far western animals. Eye has dark grey, comma-shaped mark. Rear of body gradually becomes lighter. Ventral colour pale yellow. MB 15 rows, 177–207 VENT, SUB 69–99 and anal divided. **DISTRIBUTION** From Borroloola, south to Yulara in NT, north-west to Halls Creek, WA. A presumably isolated population in eastern Pilbara between Little Sandy Desert and Nifty Mine. **HABITAT AND HABITS** Lives in tropical savannah, spinifex-dominated woodland and sand-ridge deserts. Diurnal. Feeds on lizards. Probably lays eggs. Bites may cause significant swelling and pain. HARMFUL

Grey Whip Snake ■ *Demansia simplex* TL 54cm

DESCRIPTION Top of body plain grey. Vertebral zone slightly darker than flanks. Head grey to tan. Yellow stripe runs along each of lower flanks. Eye has white-edged, dark grey, comma-shaped mark. Belly white to cream. More robust than other whip snakes. MB 15 rows, 140–150 VENT, SUB 55–65 and anal divided. **DISTRIBUTION** WA, from Kimberley region, across northern section of NT to Jabiru. **HABITAT AND HABITS** Lives in rocky woodland and tropical savannah. Diurnal. Feeds on lizards. Probably lays eggs. Bites may cause significant swelling and pain, but may result in more serious symptoms. HARMFUL

Collared Whip Snake ■ *Demansia torquata* TL 70cm

DESCRIPTION Top of body brown to bluish-grey. Head and nape grey with pale sides to head and 2 thin, pale yellow to orange bands. Eye has dark grey, comma-shaped mark. Rear of body gradually becomes lighter, particularly in young individuals. Ventral colour reddish. Young individuals brighter in colouration than adults. MB 15 rows, 185–220 VENT, SUB 70–90 and anal divided. **DISTRIBUTION** Eastern Qld, from Gladstone and tip of Cape York Peninsula. **HABITAT AND HABITS** Lives in tropical savannah and open woodland. Diurnal. Feeds on lizards. Lays 5–7 eggs. Bites may cause significant swelling and pain, but may result in more serious symptoms. VENOMOUS

Lesser Black Whip Snake ■ *Demansia vestigiata* TL 120cm

DESCRIPTION Top of body grey to black, with each dorsal scale having a darker rear edge, giving snake a variegated appearance. Some individuals have a yellow forebody.

Belly grey. MB 15 rows, 160–220 VENT, SUB 70–95 and anal divided. **DISTRIBUTION** From Ipswich, up east coast, and across north coast of Australia. Also southern PNG. **HABITAT AND HABITS** Lives in dry forests, open woodland and grassland. Diurnal. Diet consists of lizards, frogs and occasionally other snakes. Lays 3–17 eggs. Only thought to be a potentially dangerous species until the death of a young man in PNG in 2007. Polyvalent Antivenom may neutralize bites from this species. DANGEROUSLY VENOMOUS

De Vis' Banded Snake ■ *Denisonia devisi* TL 70cm

DESCRIPTION Top of body brown to grey with dark brown to black, broad cross-bands. Some populations weakly banded or blotched. Head dark brown to grey, with white-and-black barred lips. Belly opaline-white. MB 17 rows, 120–150 VENT, SUB 20–40 and anal single. **DISTRIBUTION** From Wallpolla Island, Vic, through western NSW and central

and western Qld, to Gregory. **HABITAT AND HABITS** Lives in open woodland, black-soil plains, swamps and inland river margins. Nocturnal. Feeds on frogs, lizards and occasionally small rodents. Gives birth to 3–12 young. Bites can cause significant swelling and pain, and have resulted in marked symptoms. VENOMOUS

Ornamental Snake ■ *Denisonia maculata* TL 50cm

DESCRIPTION Top of body brownish-grey with dark grey. Head dark brown to grey, usually slightly darker than body. Labial scales mottled or barred with pale grey to black.

Belly grey with dark flecking under head and forebody. Tail can be bright yellow in juveniles. MB 17 rows, 120–150 VENT, SUB 20–40 and anal single. **DISTRIBUTION** Qld, from Eidsvold through to Charters Towers. **HABITAT AND HABITS** Lives in open woodland, around swamp and inland river margins, usually in deep cracking soils. Nocturnal. Feeds on frogs and occasionally lizards. Gives birth to 3–11 young. Bites can cause significant swelling and pain, and have resulted in marked symptoms. VENOMOUS

White-lipped Snake ■ *Drysdalia coronoides* TL 40cm

DESCRIPTION Top of body highly variable, from tan to reddish, brown, grey or greenish. Prominent ragged white stripe along lips usually bordered with dark red, brown or black.

Belly yellow to orange-red. MB 15 rows, 120–160 VENT, SUB 35–70 and anal single. **DISTRIBUTION** Far south-east SA, through Tas and southern Vic, and highland areas in eastern NSW. **HABITAT AND HABITS** Lives in grassland, heaths and forests. Often basks in grass tussocks and on vegetation. Diurnal. Feeds on lizards and occasionally frogs. Gives birth to 2–10 young. Bites can cause significant swelling and pain, and have resulted in marked symptoms. VENOMOUS

Masters' Snake ■ *Drysdalia masterii* TL 40cm

DESCRIPTION Top of body brownish to grey, often with fine dark flecking. Head dark brown to grey, usually slightly darker than body. Pale cream to orange or pale brown collar on nape. Labial scales mottled or barred with pale grey to black, often with fine black stripe. Belly orange to yellow. MB 15 rows, 140–160 VENT, SUB 40–55 and anal single.

DISTRIBUTION Southern Australia in three separate populations: 1. extends from Big Desert, Vic, to Tailem Bend, SA; 2. on southern Yorke Peninsula; 3. in coastal regions of Eyre Peninsula across to Esperance, WA. **HABITAT AND HABITS** Lives in open woodland, mallee, heaths and chenopod shrubland. Diurnal. Feeds on lizards. Gives birth to 2–4 young. HARMFUL

Mustard-bellied Snake ■ *Drysdalia rhodogaster* TL 50cm
(Eastern Masters' Snake)

DESCRIPTION Top of body brownish to grey, often with fine dark flecking. Head dark brown to grey, usually slightly darker than body. Pale cream to orange or pale brown collar on nape. Labial scales mottled or barred with pale grey to black, often with fine black stripe. Belly yellow to pink. MB 15 rows, 140–160 VENT, SUB 40–55 and anal single.

DISTRIBUTION NSW, from central coast region, south to Vic border east of GDR. **HABITAT AND HABITS** Lives in woodland, grassland and heaths. Basks sitting in centres of grass tussocks. Diurnal. Feeds on lizards. Gives birth to 2–6 young. Bites can cause significant swelling and pain, and have resulted in marked symptoms. HARMFUL

Bardick ■ *Echiopsis curta* TL 70cm

DESCRIPTION Top of body yellow to dark brown or grey. Some individuals reddish-orange, and some have barring on lips. Belly yellowish to cream. MB 17–21 (usually 19) rows, 120–155 VENT, SUB 25–40 and anal single. **DISTRIBUTION** Disjunct distribution across southern Australia, with three separate populations: 1. in south-west WA; 2. on Eyre Peninsula, SA; 3. in western Vic, south-west NSW and neighbouring SA. **HABITAT AND HABITS** Lives in open woodland, mallee, heaths and coastal dune assemblages. Shelters under cover, for example beneath logs, in clumps of spinifex and beneath man-made debris. Mainly nocturnal. Feeds on lizards, frogs, small mammals, and occasionally birds and insects. Gives birth to 3–14 young. Death Adder or Polyvalent Antivenom is used to neutralize bites from this species. DANGEROUSLY VENOMOUS

Western Crowned Snake ■ *Elapognathus coronatus* TL 70cm
(Werr)

DESCRIPTION Top of body tan, brown, olive to grey. Head silver to bluish-grey, bordered with black band across back of head. Labial scales bisected by black upper-edged, white stripe. Belly reddish-orange to yellow. MB 15 rows, 130–160 VENT, SUB 35–50 and anal single. **DISTRIBUTION** South-west WA, from Muchea south along coast, and east along Great Australian Bight to Point Culver. Also on Archipelago of Recherche. **HABITAT AND HABITS** Lives in open woodland, swamps and heaths. Diurnal. Feeds on lizards, and occasionally frogs and rodents. Gives birth to 3–9 young. VENOMOUS

Short-nosed Snake ■ *Elapognathus minor* TL 50cm
(Little Brown Snake)

DESCRIPTION Top of body brownish to grey. Skin between scales black, giving reticulated appearance when inflated. Lower flanks can be reddish. Eye large and short snout, as its name suggests. Lips light grey with dark margins on rear of each supralabial. Dark band extends from nape through pale lateral margin behind head. Belly greenish-yellow with orange-tinted stripes on lateral edges. MB 15 rows, 115–130 VENT, SUB 40–55 and anal single. **DISTRIBUTION** South-west WA, from Busselton, across to Two People's Bay. **HABITAT AND HABITS** Lives in closed woodland, swamps and wet heaths. Diurnal. Feeds on lizards. Gives birth to 8–12 young. VENOMOUS

Turtle-headed Sea Snake ■ *Emydocephalus annulatus* TL 90cm

DESCRIPTION Very variable. Some individuals heavily patterned, others plain. Usually dark coloured with light markings forming bands and rings. Head blunt and rounded. Males have conical-shaped projection on end of rostral scale. MB 15–17 rows, 128–144 VENT, SUB 24–30 all single and anal single. **DISTRIBUTION** Waters from Barrow Island to Ashmore Reef, WA, in Qld from Gladstone to Brisbane, and vagrant individuals off Sydney, NSW. Also tropical waters between Timor and New Caledonia. **HABITAT AND HABITS** Lives in shallow waters over coral reefs. Diurnal. Known to only eat fish eggs. Gives birth to 2–8 young. Potentially a species complex. Venom apparatus has atrophied due to dietary specialization. HARMFUL

Western population

Courting pair

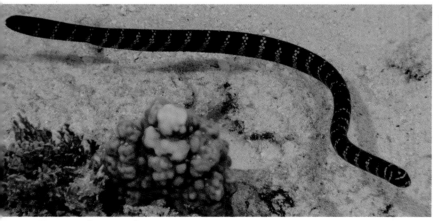

Eastern population

Mangrove Sea Snake ■ *Ephalophis greyae* TL 65cm

DESCRIPTION Top of body cream to white with grey to black bands. Skin between scales black, forming reticulated pattern. Belly colour similar to upper surface. Juveniles brighter than adults, with distinct markings. MB 19–21 rows, 151–184 VENT, SUB 24–33 all single and anal divided. **DISTRIBUTION** Waters from Shark Bay to Kings Sound, WA. **HABITAT AND HABITS** Lives in tidal waters of estuaries and mangrove-lined creeks. Activity period dictated by tidal influences. Known to eat small, hole-dwelling fish. Gives birth to 3–6 live young. VENOMOUS

Yellow-naped Snake ■ *Furina barnardi* TL 50cm

DESCRIPTION Top of body pale to dark brown, grey or almost black. Most individuals have thin, pale yellow to orange band across nape, which is absent in many older individuals. Head and nape chocolate-brown to black, and supralabials usually pale. Belly white. MB 17 rows, 170–200 VENT, SUB 35–50 and anal divided. Nasal scale usually divided, while in the Orange-naped Snake (see p. 75) it is entire. **DISTRIBUTION** Qld, from Mareeba south to Gladstone, and inland to around Barcaldine. **HABITAT AND HABITS** Lives in open woodland, brigalow and savannah. Nocturnal. Feeds on lizards. Lays 7–10 eggs. Possibly a colour variation of the Orange-naped Snake. VENOMOUS

Red-naped Snake ■ *Furina diadema* TL 40cm

DESCRIPTION Top of body reddish to dark brown or brownish-grey. Edges of midbody scales darker, forming reticulated appearance. Most individuals have an orange to reddish mark enclosed within black nape. Head and nape black, and supralabials usually pale. Belly white. MB 15 rows, 160–210 VENT, SUB 35–70 and anal divided. Nape-band bisects the black in the Orange-naped Snake (see opposite), but remains within the black in this species. **DISTRIBUTION** Qld, from Rockhampton south and west, covering most of NSW, eastern SA including Flinders Ranges to Port Augusta, and far north-west Vic. **HABITAT AND HABITS** Lives in open woodland, mulga, brigalow, sand-ridge deserts, rocky outcrops and mallee. Nocturnal. Feeds on lizards. Lays 2–5 eggs. HARMFUL

Orange-naped Snake ■ *Furina ornata* TL 70cm
(Moon Snake)

DESCRIPTION Top of body reddish to dark brown, grey or almost black; edges of midbody scales darker, forming reticulated appearance. Most individuals have a broad orange to reddish band across nape. This is indistinct in some older females. Head and nape black, and supralabials usually pale. Belly white. MB 15–17 rows, 160–240 VENT, SUB 35–70 and anal divided. Nasal scale entire, while in the Yellow-naped Snake (see p. 73) it is usually divided. Midbody scales in 15 rows in southern WA and parts of Qld.

DISTRIBUTION Across northern and central Australia from north of goldfields, WA, across through north-west SA,

all of NT, western and northern Qld, reaching coast at Bowen. Also Torres Strait Islands.

HABITAT AND HABITS Lives in open woodland, mulga, deserts, black-soil plains, gorges, brigalow and savannah. Nocturnal. Feeds on lizards. Lays 3–6 eggs. Possibly a species complex – northern animals are larger than southern ones. VENOMOUS

Dunmall's Snake ■ *Glyphodon dunmalli* TL 75cm

DESCRIPTION Top of body dark brown to greyish-black. Occasionally some yellowish flushing on supralabials and rear of head. Belly white to light grey. MB 21 rows, 175–190 VENT, SUB 35–50 and anal divided. Nasal scale divided. **DISTRIBUTION** From NSW–Qld border, north around Rockhampton, Qld. **HABITAT AND HABITS** Lives in dry forests, brigalow and scrubland, where it can be seen foraging on the ground or crossing roads. Shelters under cover, for example beneath logs and rocks. Nocturnal. Diet comprises lizards, reptile eggs and frogs. Lays 5–9 eggs. Sometimes placed in same genus as the naped snakes *Furina*, to which it is related. No known antivenom is currently recommended for bites from this species. VENOMOUS

Brown-headed Snake ■ *Glyphodon tristis* TL 100cm

DESCRIPTION Top of body brown to greyish-black. Conspicuous yellow band across nape that fades to obscurity in old individuals. Head usually lighter brown than body. White skin between scales prominent, giving reticulated appearance. Belly white to light grey. MB 19 rows, 160–190 VENT, SUB 30–60 and anal divided. Nasal scale divided. **DISTRIBUTION** Qld, north of Edmonton. Also PNG. Suggested by some to occur in Arnhem Land, NT, but no specimens are recorded. **HABITAT AND HABITS** Lives in forests, vine thickets, savannah and scrubland. Often found foraging on the forest floor or crossing roads. Shelters under cover, for example beneath logs and rocks. Nocturnal. Diet comprises lizards, reptile eggs and frogs. Lays 6–8 eggs. Sometimes placed in same genus as the naped snakes *Furina*, to which it is related. Polyvalent Antivenom has been used for bites from this species. VENOMOUS

Grey Snake ■ *Hemiaspis damelii* TL 55cm

DESCRIPTION Top of body tan to brown, or grey. Nape has distinct black band that can extend on to top of head. Belly cream to light grey. MB 17 rows, 140–170 VENT, SUB 35–50 single and anal divided.
DISTRIBUTION Qld, from Rockhampton, through southern brigalow belt, to Macquarie Marshes. Isolated population in south-central NSW, around Waugorah and Balranald. **HABITAT AND HABITS** Lives in grassland, brigalow and open woodland, mainly around river systems. Crepuscular to nocturnal. Feeds on frogs and occasionally lizards. Gives birth to 4–16 young. Bites can cause significant swelling and pain, and have resulted in marked symptoms. VENOMOUS

Marsh Snake ■ *Hemiaspis signata* TL 90cm
(Black-bellied Swamp Snake; Swamp Snake)

DESCRIPTION Top of body highly variable, from all shades of brown, greenish and grey, to black. Head can be olive, becoming yellowish lower down towards ventral surface. Prominent ragged white stripes along lips, extending behind eye on to neck. Belly black to grey. MB 17 rows, 150–170 VENT, SUB 40–60 single and anal divided. **DISTRIBUTION** Isolated populations centred on Cairns and Eungella, northern Qld. Further south extends from Gladstone, Qld, along GDR, to Nowra, NSW. **HABITAT AND HABITS** Lives in rainforests, grassland, heaths and swamps. Often seen basking in grass tussocks and on vegetation. Diurnal. Feeds on frogs and occasionally lizards. Gives birth to 4–16 young. Bites can cause significant swelling and pain, and have resulted in marked symptoms. VENOMOUS

Pale-headed Snake ■ *Hoplocephalus bitorquatus* TL 60cm

DESCRIPTION Top of body silver-grey to black. Conspicuous white band across nape on juveniles and young individuals that fades to pale saddle in old individuals. Face marked with black spots. Belly grey. MB 19–21 rows, 190–225 VENT, SUB 40–65 and anal single. **DISTRIBUTION** Disjunct, with one population in northern Qld and another

further south in southern Qld into NSW. **HABITAT AND HABITS** Lives in open woodland, brigalow and scrubland. Often found sitting exposed on trunks of large trees or crossing roads. Shelters under cover, for example beneath logs, in tree hollows and under bark of standing trees. Nocturnal. Diet comprises lizards, frogs and small mammals. Gives birth to 2–17 young. Toxic venom has caused serious symptoms rapidly following a bite. Tiger Snake or Polyvalent Antivenom is used to neutralize bites from this species. DANGEROUSLY VENOMOUS

Broad-headed Snake ■ *Hoplocephalus bungaroides* TL 90cm

DESCRIPTION Top of body black with yellow to white flecks that form irregular, thin cross-bands. Belly grey. MB 21 rows, 200–230 VENT, SUB 40–65 and anal single. **DISTRIBUTION** Sydney and surrounding areas of NSW. **HABITAT AND HABITS** Lives

in dry forests and heaths on sandstone escarpments with rock exfoliations. Occasionally seen partially exposed by day basking, or while moving on rock faces at night. Shelters under cover, for example beneath rocks and behind loose bark of standing trees. Nocturnal. Diet comprises lizards and occasionally small mammals. Gives birth to 2–12 young. Reported to have caused a fatality, but this has not been substantiated. Vulnerable; threatened by habitat destruction and fragmentation. Tiger Snake or Polyvalent Antivenom is used to neutralize bites from this species. DANGEROUSLY VENOMOUS

Stephens' Banded Snake ▪ *Hoplocephalus stephensii* TL 110cm

DESCRIPTION Top of body dark grey to black. Bands more distinct in young and juveniles than in adults, and unbanded populations are known, which can resemble the Pale-headed Snake (see opposite). Face blotched with black spots; supralabials marked with black and white. Belly grey. MB 21 rows, 225–250 VENT, SUB 50–70 and anal single. **DISTRIBUTION** Maryborough, Qld, to Ourimbah, NSW. Two isolated populations further north at Eungella and Kroombit Tops, Qld. **HABITAT AND HABITS** Lives in forests, granite outcrops and scrubland. Shelters under cover, for example beneath logs, in tree hollows and under bark of standing trees. Nocturnal. Diet comprises lizards, frogs and small mammals. Gives birth to 2–17 young. Toxic venom has caused serious symptoms rapidly following a bite. Tiger Snake or Polyvalent Antivenom is used to neutralize bites from this species. DANGEROUSLY VENOMOUS

'Unbanded' race

Black-ringed Mangrove Snake ■ *Hydrelaps darwiniensis* TL 52cm

DESCRIPTION Top of body cream to yellow with black bands. Head sometimes grey or

black. Belly colour similar to upper surface. MB 25–27 rows, 160–179 VENT, SUB 20–36 all single and anal divided. **DISTRIBUTION** Waters from Dampier, WA, across northern Australia to western Cape York Peninsula, Qld. Also southern New Guinea. **HABITAT AND HABITS** Lives in shallow tidal waters of estuaries and mangrove-lined creeks. Nocturnal. Known to eat small, hole-dwelling fish. Gives birth to 3–6 young. VENOMOUS

Black-headed Sea Snake ■ *Hydrophis atriceps* TL 100cm

DESCRIPTION Top of body cream to white, with grey to black bands. Belly colour similar to upper surface. Head and nuchal region black, as its name suggests. Juveniles much brighter than adults, with clear black markings. MB 39–49 rows, 371–392 VENT,

SUB 47–59 all single and anal divided. **DISTRIBUTION** Waters around northwest Australia from Darwin, NT. Also Indonesia and southern PNG. **HABITAT AND HABITS** Lives in waters over coral reefs and rocky areas. Nocturnal. Known to eat small eels and hole-dwelling fish. Gives birth to 1–7 young. Sea Snake Antivenom is expected to neutralize bites from this species. DANGEROUSLY VENOMOUS

Belcher's Sea Snake ■ *Hydrophis belcheri* TL 100cm

DESCRIPTION Top of body cream to beige with grey, poorly defined bands. Belly colour similar to upper surface. Juveniles much brighter than adults, with clear dark markings. MB 32–36 rows, 278–313 VENT, SUB 28–43 all single and anal divided. **DISTRIBUTION** Waters north of Australia, with waifs entering Arafura Sea from Indonesia and PNG. **HABITAT AND HABITS** Lives in waters over coral reefs. Nocturnal. Known to eat small eels. Gives birth to 2–4 young. Often incorrectly suggested as the most toxic snake in the world to mice. Sea Snake Antivenom is expected to neutralize bites from this species. DANGEROUSLY VENOMOUS

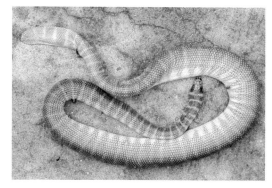

Dwarf Sea Snake ■ *Hydrophis caerulescens* TL 80cm

DESCRIPTION Top of body cream to bluish-grey with darker bands. Belly colour similar to upper surface. Juveniles brighter than adults, with clear dark markings. MB 38–54 rows, 253–334 'VENT' SUB 40–52 all single and anal divided. Differs from the Rough-scaled Sea Snake (see p. 84) by having higher MB count. **DISTRIBUTION** Waters off northern Australia, in Gulf of Carpentaria, Qld, and a single record from Fitzroy River, eastern Qld. Also throughout Southeast Asia. **HABITAT AND HABITS** Lives in river mouths, mangroves and estuaries. Nocturnal. Known to eat small eels and hole-dwelling fish. Gives birth to 2–15 young. Sea Snake Antivenom is expected to neutralize bites from this species. DANGEROUSLY VENOMOUS

Slender-necked Sea Snake ▪ *Hydrophis coggeri* TL 120cm
(Cogger's Sea Snake)

DESCRIPTION Top of body cream to brown, with dark grey to black bands. Belly colour similar to upper surface. Juveniles much brighter than adults, with clear black markings. MB 34–35 rows, 298–322 VENT, SUB 37 all single and anal divided. **DISTRIBUTION** Waters off north-west Australia, at Ashmore and Scott's Reefs. Also Timor Sea, Vanuatu, New Caledonia and Fiji. **HABITAT AND HABITS** Lives in waters over coral reefs and rocky areas. Also hunts over sand flats. Nocturnal. Known to eat snake eels. Gives birth to 3–8 young. Sea Snake Antivenom is expected to neutralize bites from this species. DANGEROUSLY VENOMOUS

Spine-bellied Sea Snake ■ *Hydrophis curtus* TL 100cm

DESCRIPTION Top of body pale brown to grey above, with wide cream to white bands that fade with age. Juveniles white or cream with grey bands. Belly colour similar to upper surface. MB 25–43 rows, 114–230 VENT, SUB 239–43 all single and anal divided.

DISTRIBUTION North of Gladstone, Qld, to Kimberleys, WA. Also across Indian Ocean into Arabian Gulf. **HABITAT AND HABITS** Lives in waters over coral reefs, rocky areas and sandy estuaries. Sexually dimorphic, with males developing spine-like projections on lower scales. Mainly nocturnal. Known to eat a wide variety of fish. Gives birth to 1–15 young. Sea Snake Antivenom is used to neutralize bites from this species. DANGEROUSLY VENOMOUS

Juvenile

Geometrical Sea Snake ■ *Hydrophis czeblukovi* TL 120cm

DESCRIPTION Top of body light to dark grey, with large, narrow, cream to pale yellow hexagonal markings. Belly usually grey. Juveniles thought to be similar to adults. MB

55–56 rows, 315–324 VENT, SUB 48–52 all single and anal divided. **DISTRIBUTION** Waters from Shark Bay, WA, up northern coast to Arnhem Land, NT. Also Indonesia and PNG. **HABITAT AND HABITS** Lives in waters over coral reefs and sandy areas. Nocturnal. Known to eat eels. Probably gives birth to live young. Sea Snake Antivenom is expected to neutralize bites from this species. DANGEROUSLY VENOMOUS

Rough-scaled Sea Snake ■ *Hydrophis donaldi* TL 90cm

DESCRIPTION Top of body light brown to gold, with or without pale grey-brown bands. Belly colour similar to upper surface. Juveniles thought to be similar to adults. MB 33–35 rows, 246–288 VENT, SUB 42–51 all single and anal divided. Differs from the Dwarf

Sea Snake (see p. 81) by having lower MB count. **DISTRIBUTION** Waters of Gulf of Carpentaria, Qld. **HABITAT AND HABITS** Lives in waters over mud flats, estuaries and shallow, protected areas, including seagrass beds. Nocturnal. Sexually dimorphic, with males developing spine-like projections on lower scales. Probably gives birth to live young. Recently described to science. Sea Snake Antivenom is expected to neutralize bites from this species. DANGEROUSLY VENOMOUS

Elegant Sea Snake ■ *Hydrophis elegans* TL 180cm

DESCRIPTION Top of body cream to white, with fine, broad grey to black bands. Belly colour similar to upper surface. Juveniles much brighter than adults, with clear black markings. MB 39–49 rows, 345–433 VENT, SUB 36–43 all single and anal divided. **DISTRIBUTION** Waters around Australia, north of Sydney, NSW, to Exmouth, WA. **HABITAT AND HABITS** Lives in waters over coral reefs and rocky areas. Commonly encountered across much of its range. Nocturnal. Known to eat small eels and hole-dwelling fish. Gives birth to 3–30 live young. One of the longest of the sea snakes. Sea Snake Antivenom is expected to neutralize bites from this species. DANGEROUSLY VENOMOUS

Plain Sea Snake ■ *Hydrophis inornatus* TL 80cm

DESCRIPTION Top of body pale grey to beige with sparse pale bands. Belly colour similar to upper surface. Juveniles thought to be similar to adults. MB 35–48 rows, 195–293

VENT, SUB 36–43 all single and anal divided. **DISTRIBUTION** Australian waters, in Torres Strait off Daru Island. Also off PNG coast. **HABITAT AND HABITS** Lives in waters over coral reefs and rocky areas. Nocturnal. Known to eat small eels and gobies. Gives birth to 4–10 live young. It has been suggested that individuals from Australia are incorrectly identified and that the species does not occur in Australian waters. Sea Snake Antivenom is expected to neutralize bites from this species. DANGEROUSLY VENOMOUS

Spectacled Sea Snake ■ *Hydrophis kingii* TL 180cm

DESCRIPTION Upper surface from head and first third usually has broad dark bands fading to bluish-grey over rear two-thirds. Pale grey to cream between bands. Belly colour similar to upper surface. Usually has a white ring around the eyes. MB 37–39 rows, 324–342 VENT, SUB 36–43 all single and anal divided. **DISTRIBUTION** Waters north of Coffs

Harbour, NSW, across northern Australia to Barrow Island, WA. Also Indonesia and PNG. **HABITAT AND HABITS** Lives in waters over coral reefs and rocky areas, as well over mudflats and seagrass beds. Nocturnal. Probably feeds on eels and small fish. Gives birth to 1–8 live young. Sea Snake Antivenom is expected to neutralize bites from this species. DANGEROUSLY VENOMOUS

Laboute's Sea Snake ■ *Hydrophis laboutei* TL 100cm

DESCRIPTION Top of body black with thin cream bands. Belly colour similar to upper surface. MB 44–46 rows, 265–280 VENT, SUB 35–39 and anal divided.
DISTRIBUTION Waters east of Australia at Chesterfield Reef, and expected to occur in Australian waters. Also New Caledonia. **HABITAT AND HABITS** Lives in waters over coral reefs, to a depth of 62m. Nocturnal. Thought to feed on small eels and hole-dwelling fish. Presumed to give birth to live young. Sea Snake Antivenom is expected to neutralize bites from this species. VENOMOUS

Small-headed Sea Snake ■ *Hydrophis macdowelli* TL 80cm

DESCRIPTION Top of body pale cream to white, with grey dorsal bands above forming spots on lower flanks. Bands fade with age. Belly colour similar to upper surface. MB 35–39 rows, 252–274 VENT, SUB 36–44 all single and anal divided. **DISTRIBUTION** North of Brisbane, Qld, to Barrow Island, WA. Also PNG and New Caledonia. **HABITAT AND HABITS** Lives in waters over coral reefs, rocky areas and sandy estuaries. Mainly nocturnal. Known to feed on a wide variety of fish. Gives birth to 2 live young. Sea Snake Antivenom is used to neutralize bites from this species. DANGEROUSLY VENOMOUS

Olive-headed Sea Snake ■ *Hydrophis major* TL 160cm

DESCRIPTION Top of body cream to white, with grey to black, elongated cross-bars; thin dark bands in the pale inter-spaces. Belly colour similar to upper surface. Juveniles much

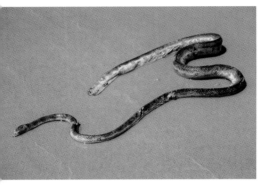

brighter than adults, with clear black markings. MB 33–43 rows, 197–245 VENT, SUB 39–43 all single and anal divided. **DISTRIBUTION** Waters around Australia, north of Sydney, NSW, to Bunbury, WA. **HABITAT AND HABITS** Lives in waters over coral reefs and rocky areas, as well as sandy inlets, estuaries and seagrass beds. Nocturnal. Known to feed on small fish. Gives birth to 6–12 live young. Sea Snake Antivenom is expected to neutralize bites from this species. DANGEROUSLY VENOMOUS

Black-banded Robust Sea Snake ■ *Hydrophis melanosoma* TL 100cm

DESCRIPTION Top of body black with cream to yellow, thin bands. Belly colour similar to upper surface. MB 37–43 rows, 260–370 VENT and anal divided. **DISTRIBUTION**

Waters north of Australia, in Torres Strait. Also throughout Indonesia and Malaysia. **HABITAT AND HABITS** Lives in waters over coral reefs, shoals and into estuaries. Nocturnal. Known to feed on eels and catfish. Gives birth to 6 young. It has been suggested that individuals from Australia are incorrectly identified and that the species does not occur in Australian waters. Sea Snake Antivenom is expected to neutralize bites from this species. DANGEROUSLY VENOMOUS

Spotted Sea Snake ▪ *Hydrophis ornatus* TL 150cm
(Ornate Sea Snake)

DESCRIPTION Top of body whitish-cream laterally with pale grey dorsum, and broad, grey-brown blotches and markings. Belly colour similar to flanks. MB 39–59 rows, 240–340 VENT, SUB 38–52 single and anal divided. **DISTRIBUTION** Waters around Australia, from Barrow Island, WA, across northern and eastern Australia, down to Brisbane, Qld. Waifs occur as far south as Tas east coast. Throughout southern Asia across to Persian Gulf. **HABITAT AND HABITS** Lives in waters over coral reefs and rocky areas, as well as over mudflats and seagrass beds. Nocturnal. Known to feed on gobies and other small fish. Gives birth to 1–6 young. Australian individuals known as *H. ocellatus* by some authorities. Sea Snake Antivenom is expected to neutralize bites from this species. DANGEROUSLY VENOMOUS

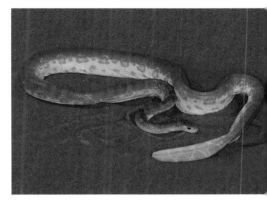

Large-headed Sea Snake ▪ *Hydrophis pacificus* TL 150cm

DESCRIPTION Top of body grey, usually with dark bands that become obscure with age. Belly pale yellow to creamy-white. Juveniles have clear black markings. Head large and distinct. MB 45–49 rows, 320–430 VENT and anal divided. **DISTRIBUTION** Waters in Gulf of Carpentaria. Also waters off PNG. **HABITAT AND HABITS** Lives in waters over coral reefs and rocky areas. Nocturnal. Known to feed on small fish. Gives birth to 17 young. The few Australian records are from prawn trawling and individuals washed up on beaches. Genetically, this species fits within Annulated Sea Snake *H. cyanocinctus* complex, and Australian animals may be split into more than one taxon in future. Sea Snake Antivenom is expected to neutralize bites from this species. DANGEROUSLY VENOMOUS

Horned Sea Snake ■ *Hydrophis peronii* TL 99cm

DESCRIPTION Top of body brown to grey with whitish blotches and markings. Some individuals banded, others plain. Belly colour similar to upper surface. MB 21–31 rows,

142–203 VENT, SUB 44 all single and anal divided. Head strongly keeled with raised scales above eyes. **DISTRIBUTION** Waters around Australia, north of Brisbane, Qld, to Barrow Island, WA. Also throughout Indonesia and the Pacific. **HABITAT AND HABITS** Lives in waters over coral reefs, mudflats and seagrass beds. Nocturnal. Known to feed on gobies and other small fish. Gives birth to 1–10 young. Sea Snake Antivenom is used to neutralize bites from this species. DANGEROUSLY VENOMOUS

Yellow-bellied Sea Snake ■ *Hydrophis platurus* TL 100cm

DESCRIPTION Top of body black to dark brown. Bottom half yellow or white. Tail white or yellow with black markings. MB 49–68 rows, 264–408 VENT, SUB 39–51 all single and anal divided. **DISTRIBUTION** From east coast of Tas, around Australian coastline from Port Phillip Bay to Perth, WA. Also throughout Indian and Pacific Oceans. **HABITAT AND HABITS** Lives in open water. Mainly diurnal. Feeds on small pelagic fish that it captures while sitting in ambush among floating debris. Gives birth to 1–6 young. The world's most widespread snake. All-yellow subspecies recently described off the coast of Costa Rica. Sea Snake Antivenom is used to neutralize bites from this species. DANGEROUSLY VENOMOUS

Stokes' Sea Snake ■ *Hydrophis stokesii* TL 150cm

DESCRIPTION Top of body white or cream, with grey to black spots that fade with age. Juveniles white or cream, with grey to black blotches. Belly colour similar to upper surface but lighter. Very heavy build. MB 54–60 rows, 252–280 VENT, SUB 33–36 all single and anal divided. **DISTRIBUTION** Waters around Australia, north of Sydney, NSW, to Exmouth, WA. Also across Indian Ocean to Arabian Gulf, and down African coast to South Africa. **HABITAT AND HABITS** Lives in waters over sandy inlets, estuaries and on coral reefs. Mainly nocturnal. Known to feed on frogfish and stonefish. Gives birth to 1–14 young. Sea Snake Antivenom is used to neutralize bites from this species. DANGEROUSLY VENOMOUS

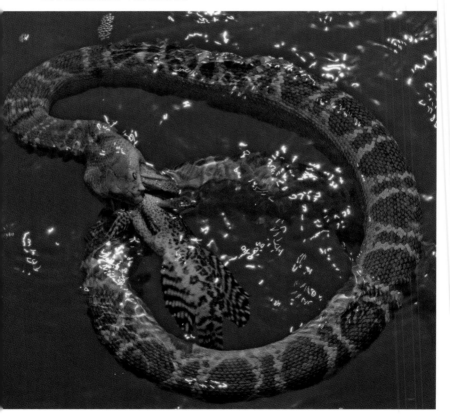

Plain-banded Sea Snake ▪ *Hydrophis vorisi* TL 60cm

DESCRIPTION Top of body cream to grey, with dark grey to black bands. Belly colour similar to upper surface. MB 29–35 rows, 330–350 VENT, SUB 36 single and anal divided. **DISTRIBUTION** Waters north of Australia, in Torres Strait. Also throughout Indonesia and Malaysia. **HABITAT AND HABITS** Lives in waters over coral reefs and rocky areas. Nocturnal. Known to feed on small eels and hole-dwelling fish. Thought to give birth to live young. Sea Snake Antivenom is expected to neutralize bites from this species. DANGEROUSLY VENOMOUS

Australian Beaked Sea Snake ▪ *Hydrophis zweifeli* TL 90cm

DESCRIPTION Top of body white or cream, with grey bands that fade with age. Juveniles white or cream, with grey to black bands. Belly colour similar to upper surface but lighter. MB 48–55 rows, 272–322 VENT, SUB 52 all single and anal divided. **DISTRIBUTION** Waters around Australia, north of Brisbane, Qld, around to about Darwin, NT. Also PNG. **HABITAT AND HABITS** Lives in waters over sandy inlets and estuaries, and occasionally found well up river in fresh water. Mainly nocturnal. Known to feed on catfish and pufferfish, and sometimes prawns. Loose skin beneath jaw may be adaption to enable it to eat larger prey. Gives birth to 1–19 young. Australian animals were originally thought to be the same as *H. schistosa*, found throughout southern Asia. Sea Snake Antivenom is used to neutralize bites from this species. DANGEROUSLY VENOMOUS

White-lipped Sea Krait ■ *Laticauda colubrina* TL 160cm

DESCRIPTION Usually blue to blue-grey with black bands. Lower lateral surfaces lighter, before becoming cream to yellow underneath. MB 21–25 rows, 210–250 VENT, SUB

25–50 all divided and anal divided. Differs from the Black-lipped Sea Krait (see below) by pale lip colouration and higher midbody scale row count. **DISTRIBUTION** Across northern Australia, with records from Darwin, NT, and in Torres Strait, but all individuals found are better described as waifs. Also southern Asia into Pacific across to Tonga. **HABITAT AND HABITS** Aquatic, but readily found on land outside Australia, entering forests, gorges and caves, and climbing into trees and on to cliff faces. Nocturnal, but active by day while hunting prey. Feeds mainly on fish and eels. Occasionally scavenges dead fish found in holes at low tide. Lays 5–19 eggs. DANGEROUSLY VENOMOUS

Black-lipped Sea Krait ■ *Laticauda laticaudata* TL 100cm

DESCRIPTION Usually blue to blue-grey with black bands. Lower lateral surfaces lighter, before becoming cream to yellow underneath. MB 19 rows, 225–245 VENT, SUB 25–50

all divided and anal divided. Differs from the White-lipped Sea Krait (see above) by dark lip colouration and lower midbody scale row count. **DISTRIBUTION** Across northern Australia, but all individuals found are better described as waifs. Also Asia into Pacific across to Tonga. **HABITAT AND HABITS** Aquatic, but readily found on land outside Australia, entering forests, rocky seaside shelves and caves. Nocturnal, but active by day while hunting food. Feeds mainly on fish and eels. Lays 5–11 eggs. DANGEROUSLY VENOMOUS

Slender Sea Snake ▪ *Microcephalophis gracilis* TL 70cm

DESCRIPTION Top of body cream to white, with distinct dark grey to black bands. Belly colour lighter than upper surface. Juveniles brighter than adults, with clear black markings. Head and neck very small. MB 30–36 rows, 220–287 VENT and anal divided. **DISTRIBUTION** Waters north of Australia, in Torres Strait. Also throughout Indonesia across to Persian Gulf. **HABITAT AND HABITS** Lives in deep, turbid waters. Nocturnal. Known to feed on eels. Gives birth to 1–6 live young. Sea Snake Antivenom neutralizes bites from this species. DANGEROUSLY VENOMOUS

Black-naped Burrowing Snake ▪ *Neelaps bimaculatus* TL 45cm

DESCRIPTION Top of body yellowish-orange to reddish-brown. Each midbody scale has darker edges, forming reticulated pattern. Black mark on snout, over eyes on head and over nape. Usually creamy-yellow between snout and eye marking. Belly cream to white. MB 15 rows, 176–228 VENT, SUB 19–30 and anal divided. **DISTRIBUTION** Eyre Peninsula, around Kingoonya, SA. Further west throughout goldfields of WA, to Bunbury and north to Shark Bay. An apparently isolated population on North West Cape. **HABITAT AND HABITS** Lives in mallee, coastal heaths and sand dunes, usually beneath the surface. Nocturnal. Feeds on small burrowing skinks. Lays 2–6 eggs. HARMLESS

Black-striped Burrowing Snake ■ *Neelaps calonotos* TL 30cm

DESCRIPTION Top of body yellowish-orange to reddish-brown. Continuous black stripe extending length of body in most individuals, but in some this is reduced to a few black vertebral spots. Each midbody scale has white spot in centre. Black mark on snout, over eyes on head and over nape. Usually creamy-yellow between the dark markings. Belly

cream to white. MB 15 rows, 126–143 VENT, SUB 23–35 and anal divided. **DISTRIBUTION** WA, from Mandurah to Lancelin, with isolated populations at Dongara and Eneabba. **HABITAT AND HABITS** Lives in coastal heaths, woodland and sand dunes, usually beneath the surface. Nocturnal. Feeds on small burrowing skinks. Lays 2–5 eggs. HARMLESS

Common Tiger Snake ■ *Notechis scutatus scutatus* TL 170cm

DESCRIPTION Very variable in both colour and pattern. Top of body can be any shade of brown, grey, black or yellow, with or without cross-bands. Belly yellow to grey. MB 17–19 rows, 140–190 VENT, SUB 35–65 and anal single. **DISTRIBUTION** From southern Qld, through eastern NSW, most of Vic, and south-east SA. **HABITAT AND HABITS** Lives in swamps, forests, grassland, rainforests, wallum, open woodland, heaths and scrubland.

Common in urban areas, often in gardens and houses. Usually diurnal, but may become nocturnal in hot weather and during summer. Diet comprises lizards, frogs, birds and small mammals. Gives birth to 5–49 young. Tiger Snake or Polyvalent Antivenom is used to neutralize bites from this species. DANGEROUSLY VENOMOUS

Krefft's Tiger Snake ■ *Notechis scutatus ater* TL 90cm

DESCRIPTION Top of body black with or without pale white, yellowish to grey crossbands. Belly blackish to grey. MB 17 rows, 163–173 VENT, SUB 41–50 and anal single.

DISTRIBUTION Southern Flinders Ranges between Crystal Brook to Mount Remarkable NP, SA. **HABITAT AND HABITS** Lives along streams in dry open forests and scrubland. Diurnal. Feeds on lizards, frogs, tadpoles, birds and small mammals. Gives birth to 6–15 young. Threatened due to habitat destruction and predation by feral species. Tiger Snake or Polyvalent Antivenom is used to neutralize bites from this species. DANGEROUSLY VENOMOUS

Tasmanian Tiger Snake ■ *Notechis scutatus humphreysi* TL 200cm

DESCRIPTION Very variable in both colour and pattern. Many different colour forms known, from whitish with grey heads, to yellow, to jet-black and all shades of brown. Some individuals are banded or speckled, while others are immaculate. Belly yellow to grey. MB 15–19 rows, 161–174 VENT, SUB 48–53 and anal single. **DISTRIBUTION** Across Tas and also on King Island group in Bass Strait. **HABITAT AND HABITS** Lives in urban environments, swamps, forests, grassland, rainforests, heaths and scrubland. Diurnal. Diet comprises lizards, frogs, birds and small mammals. Gives birth to 8–64 young. Tiger Snake or Polyvalent Antivenom is used to neutralize bites from this species. DANGEROUSLY VENOMOUS

Peninsula Tiger Snake ▪ *Notechis scutatus niger* TL 120cm

DESCRIPTION Head usually grey or black. Body coppery-brown to black. Some populations are banded with thin white to cream bands, while others are immaculate. Belly grey to black. MB 17–21 rows, 160–184 VENT, SUB 45–54 and anal single.
DISTRIBUTION SA, from Kangaroo Island, southern half of Yorke Peninsula, and along western coast of Eyre Peninsula. Also on islands in Spencer Gulf. **HABITAT AND**

HABITS Lives in swamps, forests, grassland, samphire heaths and scrubland. Diurnal. Diet comprises lizards, frogs, birds and small mammals. Gives birth to 6–38 young. Tiger Snake or Polyvalent Antivenom is used to neutralize bites from this species. DANGEROUSLY VENOMOUS

'Mainland' race

'Kangaroo Island' race

Western Tiger Snake ■ *Notechis scutatus occidentalis* TL 160cm

DESCRIPTION Top of body black or brown, with lower flanks yellow, orange or even whitish, and usually vibrant yellow cross-bands. Usually darker on rear third of body.

Belly yellow to orange with grey flecks. MB 17–19 rows, 140–165 VENT, SUB 36–51 and anal single. **DISTRIBUTION** South-west WA, from Jurien to Esperance. Also on Garden and Carnac Islands. **HABITAT AND HABITS** Lives around water courses in swamps, forests and scrubland. Often found in urban areas. Diurnal. Feeds on lizards, frogs, birds and small mammals. Gives birth to 15–35 young. Tiger Snake or Polyvalent Antivenom is used to neutralize bites from this species. DANGEROUSLY VENOMOUS

Chappell Island Tiger Snake ■ *Notechis scutatus serventyi* TL 210cm

DESCRIPTION Head usually grey or black. Body yellowish, grey, dark brown to black. Some individuals are banded with thin white to cream bands, while others are immaculate.

Lower flanks can be yellow. Belly grey to black. MB 17 rows, 160–171 VENT, SUB 47–52 and anal single. **DISTRIBUTION** Furneaux group of islands, including Flinders and Chappell Islands. **HABITAT AND HABITS** Lives in swamps, grassland, heaths and scrubland. Diurnal. Adults feed mainly on mutton-bird chicks, juveniles on small lizards and frogs. Gives birth to 6–38 young. The largest of all tiger snakes. Tiger Snake or Polyvalent Antivenom is used to neutralize bites from this species at twice the dosage for its mainland counterparts. DANGEROUSLY VENOMOUS

Inland Taipan ■ *Oxyuranus microlepidotus* TL 200cm
(Fierce Snake; Small-scaled Snake; Western Taipan)

DESCRIPTION Variable; usually yellow to chocolate-brown with black reticulations. Head usually glossy black. Goes through seasonal colour change, being dark in winter to sandy-yellow in summer. Underside bright yellow with dark flecks. MB 23 rows, 220–251 VENT, SUB 55–70 divided and anal single. **DISTRIBUTION** Far western Qld, north-east SA, south-west corner of NT, and into western NSW. Historical records at junction of Darling and Murray Rivers, but not seen there for many years. **HABITAT AND HABITS** Inhabits black-soil plains, gibber desert, grassland and savannah. Diurnal but can be crepuscular in hot weather. Basks on edges of cracks in soil. Lives in and hunts for small mammals in these cracks. This way of life is probably why it has evolved to be the world's most toxic snake to mice. Lays 8–23 eggs. Taipan or Polyvalent Antivenom is used to neutralize bites from this species. DANGEROUSLY VENOMOUS

Winter colouration

Summer colouration

Coastal Taipan ■ *Oxyuranus scutellatus scutellatus* TL 240cm
(exceptions up to 300cm)

DESCRIPTION Very variable in colour. Top of body can be any shade of brown, grey, black or yellow; some individuals have reddish-orange stripe along spine that widens towards rear. Head usually lighter, often cream to white. Belly yellow to orange, with or without red flecking. MB 21–23 rows, 220–250 VENT, SUB 45–80 divided and anal single. The Papuan Taipan (see opposite) may be a colour variation, as there are no morphological or genetic traits separating the two 'subspecies'. **DISTRIBUTION** From Qld/NSW border region across northern coastline into Kimberley region, WA. **HABITAT AND HABITS**

Lives in dry forests, grassland, savannah, open woodland, heaths and rainforest verges. Very alert, and seemingly quite uncommon in southern part of its range. Usually diurnal, but becomes nocturnal during warm weather. Diet comprises small mammals and rarely birds. Lays 5–22 eggs. Australia's, if not the world's, most dangerous snake, with very toxic venom, long fangs and nervous disposition; it can be formidable. Taipan or Polyvalent Antivenom is used to neutralize bites from this species. DANGEROUSLY VENOMOUS

Papuan Taipan ■ *Oxyuranus scutellatus canni* TL 240cm
(exceptions up to 300cm)

DESCRIPTION Very variable in colour. Top of body dark brown, grey or black, usually with reddish-orange stripe along spine that widens towards rear. Head usually lighter, often cream to white. Belly yellow to orange, with or without red flecking. MB 21–23 rows, 220–250 VENT, SUB 45–80 divided and anal single. **DISTRIBUTION** Islands on Torres Strait and PNG. Also reported from far north Qld, on Cape York Peninsula. Possibly just a colour variation of the Coastal Taipan (see opposite), as there are no morphological or genetic traits separating the two 'subspecies'. **HABITAT AND HABITS** Lives in dry forests, grassland, savannah and open woodland. Very alert, but shy unless provoked. Usually diurnal but becomes nocturnal during warm weather. Diet comprises small mammals and rarely birds. Lays 5–17 eggs. Very toxic venom, long fangs and nervous disposition. Taipan or Polyvalent Antivenom is used to neutralize bites from this species. DANGEROUSLY VENOMOUS

Western Desert Taipan ■ *Oxyuranus temporalis* TL 200cm
(Central Ranges Taipan)

DESCRIPTION Very variable in colour. From the few individuals known, yellow to dark brown. As in the Inland Taipan (see p. 101), there is a shift in individual colouration during the seasons, the colouration being lighter in summer. Belly yellow to cream, with or without orange flecking. MB 21 rows, 240–252 VENT, SUB 56–61 divided and anal single. **DISTRIBUTION** Far eastern WA, from Ilkurlka and Walter James Range, across to near Kings Canyon and George Gill Range, NT. **HABITAT AND HABITS** Lives in deserts and mallee regions with sand ridges, gravel beds, spinifex and shrubs. Shelter sites used unknown, but thought to be similar to those of other taipans. Diurnal. Diet comprises small mammals. Lays eggs, but clutch size not determined. Recently discovered, not much is known about this interesting species. Preliminary venom research has found that Taipan Antivenom will provide protection against bites from this species. DANGEROUSLY VENOMOUS

Northern Mangrove Snake ▪ *Parahydrophis mertoni* TL 50cm

DESCRIPTION Top of body greenish-yellow with grey-black bands. Belly colour similar to upper surface. MB 36–39 rows, 158–160 VENT, SUB 29–35 most single and anal divided.

DISTRIBUTION Waters from west of Darwin, NT, across to Gulf of Carpentaria, Qld. Also waters around Aru Islands, south of PNG. **HABITAT AND HABITS** Lives in shallow tidal waters of estuaries and mangrove-lined creeks. Nocturnal. Known to feed on small, hole-dwelling fish. Gives birth to 3 live young. VENOMOUS

Dwyer's Snake ▪ *Parasuta dwyeri* TL 40cm

DESCRIPTION Top of body brown to greyish. Head black on top, without black stripe along spine. Belly white to cream. MB 15 rows, 135–170 VENT, SUB 20–40 and anal single. Similar to Mitchell's Short-tailed Snake (see p. 106) but differs in lacking vertebral stripe. Northern animals lighter bodied than their southern counterparts. **DISTRIBUTION** Southern Qld, from Longreach, across to Rockhampton, south through inland NSW, west

of GDR into northern Vic. **HABITAT AND HABITS** Lives in brigalow, open woodland, grassland and scrubland. Nocturnal. Diet comprises lizards and occasionally frogs. Gives birth to 2–7 live young. Some herpetologists place this species within the Mallee Black-headed Snake (see p. 107). Bites are painful, causing swelling, blebs, headache and nausea, but may result in more serious symptoms. HARMFUL

Little Whip Snake ■ *Parasuta flagellum* TL 40cm

DESCRIPTION Top of body brown to greyish. Head black on top, with pale bar between nostrils and eyes. Belly white to cream. MB 17 rows, 125–150 VENT, SUB 20–40 and anal single. **DISTRIBUTION** From eastern SA, throughout Vic, into southern NSW.

HABITAT AND HABITS
Lives in dry forests, open woodland and grassland. Shelters under cover such as rocks and logs. Nocturnal. Diet comprises lizards and occasionally frogs. Gives birth to 2–11 young. Bites are painful, causing swelling, blebs, headache and nausea, but may result in more serious symptoms. A bite in 2007 caused the death of an adult. An anaphylactic reaction was probably involved.
VENOMOUS

Gould's Hooded Snake ■ *Parasuta gouldii* TL 50cm

DESCRIPTION Top of body orange-brown to greyish. Head black or dark brown on top, narrowing between nostrils and eyes. Belly white to cream. MB 15 rows, 150–180 VENT, SUB 25–40 and anal single. Similar to Mitchell's Short-tailed Snake (see p. 106), but differs by lacking vertebral stripe. **DISTRIBUTION** Endemic to south-west WA; found from Balladonia to Ajana, with isolated population in the Murchison. **HABITAT AND HABITS** Lives in heaths, open woodland, grassland and mallee. Nocturnal. Diet comprises lizards and occasionally frogs. Gives birth to 1–6 live young. Bites are painful, causing swelling, headache and nausea, but may result in more serious symptoms. HARMFUL

Monk Snake ■ *Parasuta monachus* TL 50cm

DESCRIPTION Top of body orange-brown, to orange or brick-red. In some individuals posterior edge to midbody scales is tipped with black. Head black or dark brown. Lower

flanks, lips and underside pearly-white. MB 15 rows, 150–180 VENT, SUB 21–35 and anal single. **DISTRIBUTION** From south-western NT and central SA, through arid WA, to Shark Bay and Pilbara. **HABITAT AND HABITS** Lives in mallee, sand-ridge deserts, open woodland and mulga. Nocturnal. Diet comprises lizards. Gives birth to 1–5 live young. Hamersley Range animals are 10 per cent larger than others, with an additional temporal scale. Bites are painful, causing swelling, headache and nausea, but may result in more serious symptoms. HARMFUL

Mitchell's Short-tailed Snake ■ *Parasuta nigriceps* TL 60cm
(Black-backed Snake)

DESCRIPTION Top of body brown to greyish. Head black on top, with black stripe running along spine that sometimes coalesces into lighter flanks. Belly white to cream. MB 15 rows, 145–175 VENT, SUB 18–38 and anal single. Similar to Dwyer's Snake (see p. 104), which differs by lacking vertebral stripe. **DISTRIBUTION** Southern inland

NSW and northern Vic, through Flinders Ranges and northern Eyre Peninsula. Also WA, from southern edge of Nullarbor Plain, across to Cervantes and south to Albany. **HABITAT AND HABITS** Lives on rocky hillsides, and in open woodland, dry forests, mallee and scrubland. Nocturnal. Diet comprises lizards and occasionally other snakes. Gives birth to 1–6 live young. Bites are painful, causing swelling, headache and nausea, but may result in more serious symptoms. HARMFUL

Mallee Black-headed Snake ■ *Parasuta spectabilis spectabilis* TL 40cm
(Port Lincoln Snake)

DESCRIPTION Top of body brown to greyish. Head black on top, with pale bar between nostrils and eyes. Black extends on to nape. Belly white, cream or yellow. MB 15 rows, 138–168 VENT, SUB 21–33 and anal single. The three subspecies are separated by distribution and scalation. **DISTRIBUTION** From western Vic, to Fowler's Bay, SA. **HABITAT AND HABITS** Found in mallee, open woodland, grassland and heaths. Shelters under cover such as rocks and logs. Diet comprises lizards and occasionally frogs. Gives birth to 2–5 live young. Bites are painful, causing swelling, headache and nausea, but may result in more serious symptoms. HARMFUL

Bush's Black-headed Snake ■ *Parasuta spectabilis bushi* TL 39cm

DESCRIPTION Top of body brown to greyish. Head black on top, with pale indent between nostrils and eyes, as well as patch on prefrontal scales. Pale spot past eye. Belly white to cream. MB 15 rows, 159–165 VENT, SUB 30–36 and anal single.

DISTRIBUTION Around Scadden, southern WA. **HABITAT AND HABITS** Lives on rocky hillsides, and in open woodland, dry forests, mallee and scrubland. Nocturnal. Diet thought to consist of lizards. Thought to give birth to live young. Only known from five specimens collected in the early 1980s. Bite effects unknown. HARMFUL

Nullarbor Black-headed Snake ■ *Parasuta spectabilis nullarbor* TL 40cm

DESCRIPTION Top of body brown to greyish. Head black on top, with pale indent between nostrils. Black not as extensive as in the Mallee Black-headed Snake (see p. 107).

Belly white to cream. MB 15 rows, 138–158 VENT, SUB 24–36 and anal single. **DISTRIBUTION** From north of Yalata, SA, west to Rawlinna, WA. **HABITAT AND HABITS** Lives on rocky slopes, and in open woodland, mallee and heaths. Known to shelter beneath limestone rocks and man-made debris. Nocturnal. Diet thought to consist of lizards. Thought to give birth to live young. Bite effects unknown. HARMFUL

Lake Cronin Snake ■ *Paroplocephalus atriceps* TL 70cm
(Black-headed Bardick)

DESCRIPTION Top of body silver-grey to brown, and head jet-black. Lips barred with white. Belly grey. MB 17–19 rows, 175–185 VENT, SUB 45–50 and anal single. **DISTRIBUTION** Around Lake Cronin, WA. **HABITAT AND HABITS** Lives in dry

forests, granite outcrops and adjoining habitats. Found sitting exposed on trunks of large Salmon Gums or crossing roads. Shelters under cover such as rocks. Nocturnal. Diet comprises lizards, frogs and small mammals. Livebearer, with litter numbers probably similar to those of the Pale-headed Snake (see p. 78). Toxic venom has rapidly caused serious symptoms following a bite. Polyvalent Antivenom has been used to neutralize bites from this species. VENOMOUS

Mulga Snake ▪ *Pseudechis australis* TL 200cm (exceptions up to 300cm)
(King Brown Snake)

DESCRIPTION Very variable in colour, from black, to pale yellow, to reddish-purple. Southern populations darker than northern populations. Some individuals have a variegated appearance, with anterior edge of the scale lighter than rear. Belly yellow to cream without orange flecking. Underside of tail pale orange in some individuals. MB 17 rows, 185–225 VENT, SUB 50–80 usually single with last few divided, anal divided.

DISTRIBUTION All over Australia's drier regions throughout almost all of WA, all of NT, most of SA, western NSW, and Qld west of GDR. **HABITAT AND HABITS** Found in dry woodland, deserts, mallee, agricultural land and heaths. Diurnal to nocturnal, depending on the temperature. Feeds on snakes, lizards, frogs, mammals and even roadkill. Lays 4–23 eggs. Despite the name, a member of the black snake genus. Injects the most venom of any Australian snake. Black Snake or Polyvalent Antivenom is used to neutralize bites from this species. DANGEROUSLY VENOMOUS

Mulga Snake continued

Spotted Mulga Snake ■ *Pseudechis butleri* TL 175cm
(Butler's Snake)

DESCRIPTION Black to dark brown, with cream to yellow spots. Juveniles grey with shiny black head. Belly yellow, and ventral scales can have black spots and black edges. MB 17 rows, 200–225 VENT, SUB 50–70 usually single with last few divided, anal divided.

DISTRIBUTION Goldfields region of southern inland WA. **HABITAT AND HABITS** Found in mulga and acacia woodland, mallee and agricultural land. Diurnal to nocturnal, depending on the temperature. Feeds on lizards, other snakes and small mammals. Lays 7–17 eggs. Black Snake or Polyvalent Antivenom is used to neutralize bites from this species. DANGEROUSLY VENOMOUS

Collett's Snake ■ *Pseudechis colletti* TL 200cm
(Downs' Tiger Snake)

DESCRIPTION Pale brown to black, with yellow, orange and pink cross-bands and spotting. Belly orange with or without dark brown to black flecking. Juveniles much brighter than adults, with orange and jet-black cross-bands. MB 19 rows, 215–235 VENT, SUB 45–65 usually single with last few divided, anal divided.

DISTRIBUTION Central western Qld. **HABITAT AND HABITS** Lives in black-soil plains and grassland. Often shelters in burrows, beneath man-made debris, under logs and rocks, and in crevices formed in cracking soils. Diurnal to nocturnal, depending on the temperature. Feeds on lizards, frogs, snakes and small mammals. Lays 7–18 eggs. Black Snake or Polyvalent Antivenom is used to neutralize bites from this species. DANGEROUSLY VENOMOUS

Spotted Black Snake ■ *Pseudechis guttatus* TL 180cm
(Blue-bellied Black Snake)

DESCRIPTION Very variable in colour, from jet-black to grey and occasionally pale brown. Some individuals flecked with yellow, red, orange or cream. Belly grey to yellowish with darker flecking. MB 17 rows, 175–230 VENT, SUB 45–65 usually single with last few divided, anal divided. **DISTRIBUTION** NSW and Qld, almost exclusively west of GDR.

In places such as west Brisbane, reaches over foothills into black-soil valleys, making it through the range. **HABITAT AND HABITS** Lives in dry woodland, savannah, grassland, brigalow and agricultural land. Often shelters in burrows, beneath man-made debris, under logs and rocks, and in cracks in the soil. Diurnal to nocturnal, depending on the temperature. Feeds on small mammals, frogs, lizards and other snakes. Lays 5–16 eggs. Tiger Snake, Black Snake or Polyvalent Antivenom is used to neutralize bites from this species. DANGEROUSLY VENOMOUS

Eastern Pygmy Mulga Snake ▪ *Pseudechis pailsei* TL 120cm

DESCRIPTION Pale yellow to golden brown, usually with magenta markings on nape. When threatened flares its neck, looking larger and more dangerous. Belly yellow to cream. Narrow head thought to be adapted for squeezing into crevices. MB 17 rows, 210–235 VENT, SUB 50–80 single and anal single. **DISTRIBUTION** Along Selwyn range, western Qld, from Winton to Riversleigh. **HABITAT AND HABITS** Lives in dry woodland and rocky gorges, and on hillsides.

Seeks shelter in rock crevices and tree hollows, or beneath debris, logs and rocks. Nocturnal, but has been found crossing roads in the morning. Feeds on lizards and frogs, and occasionally small mammals. Lays 5–11 eggs. For many years remained undetected to science, and thought to be a young mulga snake. Black Snake or Polyvalent Antivenom is used to neutralize bites from this species. DANGEROUSLY VENOMOUS

Papuan Black Snake ▪ *Pseudechis papuanus* TL 210cm

DESCRIPTION Black to dark grey, occasionally flushed with dark red. Belly dark grey. MB 19 (rarely 21) rows, 205–230 VENT, SUB 49–63 usually single with last few divided, anal divided. **DISTRIBUTION** Saibai and Boigu Islands in northern Torres Strait. Also throughout southern PNG. **HABITAT AND HABITS** Lives in swamps, forest edges and

moist grassland. Shelters in burrows, and beneath debris, logs and rocks. Diurnal, and occasionally nocturnal in hot weather. Feeds on frogs and lizards, and has been recorded eating small mammals and other snakes. Lays 7–18 eggs. Declining over much of its range due to habitat destruction and introduction of the Cane Toad. The most toxic of the black snakes. Black Snake or Polyvalent Antivenom is used to neutralize bites from this species. DANGEROUSLY VENOMOUS

Red-bellied Black Snake ■ *Pseudechis porphyriacus* TL 200cm
(Common Black Snake)

DESCRIPTION Jet-black above with orange, red or maroon markings along lower flanks. Some individuals marked with white. Head, in particular around snout, can be brown in some locations. Belly red marked with black bands. Tail black beneath. MB 17 rows, 175–215 VENT, SUB 40–65 usually single with last few divided, anal divided. Similar to small-eyed snakes (see pp. 58 and 59), differing by having clear black bands across belly. **DISTRIBUTION** Along east coast from Adelaide hills to around Cooktown, north Qld. **HABITAT AND HABITS** Inhabits swamps, creeks, forests, rainforests, urban

environments and grassland, particularly around waterbodies. Shelters in burrows or under logs, debris and rocks. Diurnal, and occasionally nocturnal during hot weather. Feeds on frogs and lizards, other snakes, fish and occasionally small mammals. Produces 5–23 young. Has declined in Qld following introduction of the Cane Toad, but appears to be beginning to recover. Tiger Snake, Black Snake or Polyvalent Antivenom is used to neutralize bites from this species. DANGEROUSLY VENOMOUS

'Northern' race

Western Pygmy Mulga Snake ■ *Pseudechis weigeli* TL 120cm

DESCRIPTION Pale yellow to golden brown or grey, with magenta markings on nape that can extend all the way down body. These reticulations are usually more prominent in western population. Belly yellow to cream. Narrow head thought to be adapted for squeezing into crevices. MB 17 rows, 210–230 VENT, SUB 50–80 single and anal single. **DISTRIBUTION** Kimberley region, WA, and Top End, NT. Disjunct population in Pilbara region, WA, almost certainly a new, as yet undescribed species. **HABITAT AND HABITS**

Lives in dry woodland, grassland, rocky gorges and hillsides. Nocturnal. Feeds on lizards and frogs, and occasionally small mammals. Lays 7–12 eggs. For many years undetected to science and thought to be a young mulga snake. Possibly comprises three species. Two are genetically distinct but require further research to determine validity of the taxa. Black Snake or Polyvalent Antivenom is used to neutralize bites from this species. DANGEROUSLY VENOMOUS

Top End population

Kimberley population

Dugite ■ *Pseudonaja affinis affinis* TL 200cm
(Spotted Brown Snake)

DESCRIPTION Very variable, from pale cream to almost black. Usually a shade of brown with random dark spots that sometimes coalesce into blotches. Head can be darker or lighter than body. Juvenile and immature individuals have black head and black band on nape. Belly yellow, to cream, to light brown, with orange-red spots and blotches. MB 19–21 rows, 190–230 VENT, SUB 50–70 and anal divided. Eastern animals have 17 MB rows, and differ from Peninsula Brown Snakes (see p. 119) by having a light ventral surface.
DISTRIBUTION Southern WA, across southern coast to western edge of Eyre Peninsula, SA. **HABITAT AND HABITS** Lives in open woodland, coastal dune associations, heathland, grassland and urban

areas. Seeks shelter in vegetation, beneath debris, or under logs and rocks. Usually diurnal, but active at night during hot weather. Feeds on small mammals and reptiles as adult; juveniles almost exclusive reptile predators. Lays 3–31 eggs. Bites from juveniles have killed healthy adult humans. Brown Snake or Polyvalent Antivenom is used to neutralize bites from this species. DANGEROUSLY VENOMOUS

Rottnest Island Dugite ■ *Pseudonaja affinis exilis* TL 130cm

DESCRIPTION Dark brown to black above and below. Juvenile and immature individuals have black head and black band on nape. MB 19 rows, 207–219 VENT, SUB 48–57 and anal divided. **DISTRIBUTION** Restricted to Rottnest Island, WA. **HABITAT AND HABITS** Lives in dry woodland, coastal dune associations and heathland. Seeks shelter in vegetation, beneath debris, or under logs and rocks. Diurnal. Feeds on small mammals and reptiles as adult; juveniles almost exclusive reptile predators. Thought to be egg layer. Brown Snake or Polyvalent Antivenom is used to neutralize bites from this species. DANGEROUSLY VENOMOUS

Tanner's Brown Snake ■ *Pseudonaja affinis tanneri* TL 120cm

DESCRIPTION Chestnut-brown to dark brown above. Ventral surface lighter with dark brown spotting. Juvenile and immature individuals have black head and black band on nape. MB 19 rows, 208–212 VENT, SUB 56–60 and anal divided. **DISTRIBUTION** Restricted to Boxer and Figure of Eight Islands in Archipelago of Recherche, off southern WA. **HABITAT AND HABITS** Lives in dry woodland and heathland. Seeks shelter in vegetation or beneath rocks. Diurnal. Feeds on small mammals and reptiles as adult; juveniles almost exclusive reptile predators. Lays 12–15 eggs. Brown Snake or Polyvalent Antivenom is used to neutralize bites from this species. DANGEROUSLY VENOMOUS

Shield-snouted Brown Snake ▪ *Pseudonaja aspidorhyncha* TL 175cm

DESCRIPTION Very variable, from pale cream to dark brown. Usually a shade of brown with or without dark bands. Almost always a black speck on nape. Juvenile and immature individuals have black head and black band on nape. Belly yellow with orange-red spots and blotches. MB 17–19 rows, 200–230 VENT, SUB 45–70 and anal divided. **DISTRIBUTION** Inland eastern Australia, including eastern SA, north-west

Vic, western NSW and southern west Qld. **HABITAT AND HABITS** Lives in open woodland, brigalow, mallee and desert, and around agricultural areas. Seeks shelter in vegetation, beneath debris, or under logs and rocks. Diurnal to nocturnal, depending on the temperature. Feeds on small mammals and reptiles as adult; juveniles almost exclusive reptile predators. Lays 9–14 eggs. Bites from juveniles have killed healthy adult humans. Brown Snake or Polyvalent Antivenom is used to neutralize bites from this species. DANGEROUSLY VENOMOUS

Speckled Brown Snake ■ *Pseudonaja guttata* TL 120cm

DESCRIPTION Very variable, but usually a shade of brown, yellow-cream or orange, with or without dark bands; many individuals flecked with black. Juveniles and immature individuals have black head and black band on nape. Belly yellow with orange-red spots and blotches. MB 19–21 rows, 190–220 VENT, SUB 45–70 and anal divided. **DISTRIBUTION** Inland northern Qld, north-east SA and NT. **HABITAT AND HABITS** Lives in grassland and black-soil plains. Seeks shelter in vegetation, beneath debris and in soil cracks. Diurnal, but occasionally crepuscular in hot weather. Feeds on frogs, lizards and occasionally small mammals; juveniles almost exclusive reptile and frog predators. Lays 3–17 eggs. Potentially a species complex. Brown Snake or Polyvalent Antivenom is used to neutralize bites from this species. DANGEROUSLY VENOMOUS

Peninsula Brown Snake ■ *Pseudonaja inframacula* TL 160cm

DESCRIPTION Very variable, from yellow-brown to purplish-black. Most individuals flecked with black, sometimes coalescing into blotches. Juveniles and immature individuals have black head and black band on nape. Belly grey. MB 17–19 rows, 190–230 VENT, SUB 50–65 and anal divided. **DISTRIBUTION** Southern SA, on Yorke and Eyre Peninsulas. Isolated population on Nullarbor Plain, WA, which has 19 midbody rows. Similar to the Dugite (see p. 116), differing by having a dark ventral surface. **HABITAT AND HABITS** Lives in open woodland, heaths, grassland and coastal dunes, and around agricultural areas. Seeks shelter in vegetation, beneath debris, or under logs and rocks. Diurnal. Feeds on small mammals, frogs and reptiles as adult; juveniles almost exclusive reptile predators. Lays 12 eggs. Brown Snake or Polyvalent Antivenom is used to neutralize bites from this species. DANGEROUSLY VENOMOUS

Ingram's Brown Snake ■ *Pseudonaja ingrami* TL 200cm

DESCRIPTION Very variable, from pale yellow, orange, reddish-brown or dark brown, to black. Juvenile and immature individuals have dark head markings. Belly yellow with

orange-red spots that are arranged in straight lines down body. MB 19–21 rows, 190–220 VENT, SUB 55–70 and anal divided. **DISTRIBUTION** Inland northern Qld and NT. **HABITAT AND HABITS** Lives in grassland and black-soil plains. Seeks shelter in soil cracks, in vegetation, and beneath logs and rocks. Diurnal. Feeds on small mammals and reptiles as adult; juveniles almost exclusive reptile predators. Lays 5–18 eggs. Brown Snake or Polyvalent Antivenom is thought to neutralize bites from this species. DANGEROUSLY VENOMOUS

Western Brown Snake ■ *Pseudonaja mengdeni* TL 140cm

DESCRIPTION Very variable, from pale cream to dark brown. Many colour morphs, including banded individuals, those marked with a herringbone pattern, and bright orange snakes with a jet-black head and neck. Juveniles and immature individuals have black head and black band on nape, while others are completely banded. Belly yellow to creamish, with orange-red spots and blotches. MB 17 rows, 190–220 VENT, SUB 55–70 and anal divided. **DISTRIBUTION** Much of arid and west Australia, including SA, western NSW, south-west Qld, arid regions of NT and most of WA, including Kimberley region. **HABITAT AND HABITS** Lives in open woodland, mallee, grassland and deserts, and around agricultural areas. Seeks shelter in vegetation, beneath man-made debris, or under logs and rocks. Diurnal, but can also be nocturnal. Feeds on small mammals and reptiles as adult; juveniles almost exclusive reptile predators. Lays 7–22 eggs. Bites from juveniles have killed healthy adult humans. Brown Snake or Polyvalent Antivenom is used to neutralize bites from this species. DANGEROUSLY VENOMOUS

Ringed Brown Snake ■ *Pseudonaja modesta* TL 60cm
(Five-ringed Snake)

DESCRIPTION Grey to reddish-brown, with 4–11 black cross-bands that fade to obscurity with age. Belly yellow with orange-red spots and blotches. MB 17 rows, 145–175 VENT, SUB 35–50 and anal divided. **DISTRIBUTION** Arid central Australia to western WA coastline. **HABITAT AND HABITS** Lives in open woodland, mallee, grassland and deserts, and around agricultural areas. Seeks shelter in vegetation, beneath man-made debris, or under logs and rocks. Cathermal. Feeds on reptiles. Lays 7–20 eggs. Recent genetic work has shown that it may not be a type of brown snake, and may be a species complex. Brown Snake or Polyvalent Antivenom is thought to neutralize bites from this species. VENOMOUS

Northern Brown Snake ■ *Pseudonaja nuchalis* TL 140cm

DESCRIPTION Very variable, from pale brown to gold or dark brown. Some individuals have dark bands, while others have dark nape-band. Usually a black speck or mark on nape. Juveniles and immature individuals have black head and black band on nape, and are sometimes completely banded. Belly yellow with orange-red spots and blotches. MB 17–19 rows, 180–230 VENT, SUB 50–70 and anal divided. **DISTRIBUTION** Across northern Australia, west of GDR into Kimberleys, WA.

HABITAT AND HABITS Lives in tropical savannah, grassland, rocky outcrops and deserts. Seeks shelter in vegetation, beneath man-made debris, in burrows, or under logs and rocks. Diurnal to nocturnal, depending on the temperature. Feeds on small mammals and reptiles as adult; juveniles almost exclusive reptile predators. Lays 8–16 eggs. Bites from juveniles have killed healthy adult humans. Brown Snake or Polyvalent Antivenom is used to neutralize bites from this species. DANGEROUSLY VENOMOUS

Eastern Brown Snake ■ *Pseudonaja textilis* TL 200cm
(Common Brown Snake)

DESCRIPTION Very variable, from pale cream to black. Usually a shade of brown with or without dark bands. Juveniles and immature individuals have black head and black band on nape, and some are completely banded. Belly yellow to cream, with orange-red spots and blotches; sometimes marked with grey. MB 17 rows, 180–235 VENT, SUB 45–75 and anal divided. **DISTRIBUTION** Eastern Australia over all of NSW, most of Vic and Qld, and south-eastern SA. Isolated populations in NT around Alice Springs, and Victoria River district, WA. Also southern PNG and Indonesia. **HABITAT AND HABITS** Inhabits woodland, brigalow, mallee, grassland, deserts and agricultural areas. Seeks shelter in vegetation, beneath man-made debris, or under logs and rocks. Diurnal to nocturnal, depending on the temperature. Feeds on small mammals and reptiles as adult; juveniles almost exclusively reptile predators. Lays 6–28 eggs. Alice Springs animals genetically similar to southern Papuan snakes and may prove to be distinct. Has killed more people than any other Australian snake, and bites from juveniles have killed healthy adult humans. Brown Snake or Polyvalent Antivenom is used to neutralize bites from this species. DANGEROUSLY VENOMOUS

Juvenile

Central Australian population

Eastern Brown Snake continued

Juvenile

Square-nosed Snake ■ *Rhinoplocephalus bicolor* TL 40cm

DESCRIPTION Top of body blackish-brown. Lower flanks pink to pale orange. Underside usually creamy-white. MB 15 rows, 135–165 VENT, SUB 20–45 and anal single.

DISTRIBUTION Southwest WA, from Esperance to Busselton. **HABITAT AND HABITS** Lives in woodland and scrubland, usually sheltering under cover, for example beneath grass trees and inside stick-ant nests. Predominantly nocturnal, but will bask on vegetation during the morning in suitable conditions. Diet comprises lizards and possibly frogs. Produces 1–5 young. HARMFUL

Desert Burrowing Snake ■ *Simoselaps anomalus* TL 41cm

DESCRIPTION Top of body yellow with regular thin, dark brown to black, straight-edged cross-bands. Head and neck background colouration white. Belly creamish-white. MB 15 rows, 115–130 VENT, SUB 15–30 and anal divided. **DISTRIBUTION** Central Australia, including north-west SA and south-western half of NT, through inland WA, reaching coast between Exmouth and Derby. **HABITAT AND HABITS** Lives on sandy soils. Shelters beneath cover and leaf litter in mulga woodland and sand-ridge deserts. Nocturnal. Feeds on ground-dwelling skinks. Lays 2–3 eggs. HARMFUL

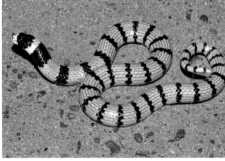

Jan's Burrowing Snake ■ *Simoselaps bertholdi* TL 32cm

DESCRIPTION Top of body orange with regular thin, dark brown to black, straight-edged cross-bands. Head and neck background colouration whitish to cream. Belly creamish-

white. MB 15 rows, 115–135 VENT, SUB 15–30 and anal divided. **DISTRIBUTION** SA from western Eyre Peninsula, across through southern Australia, to WA's western coast, south of Yannarie. **HABITAT AND HABITS** Lives on sandy soils. Shelters beneath cover and leaf litter in mallee, coastal dune assemblages, mulga woodland and sand-ridge deserts. Nocturnal. Feeds on ground-dwelling skinks. Lays 1–8 eggs. HARMFUL

West Coast Burrowing Snake ■ *Simoselaps littoralis* TL 30cm

DESCRIPTION Top of body pale yellow with regular thin, dark brown to black, straight-edged cross-bands. Head and neck background colouration whitish to cream. Belly creamish-white. MB 15 rows, 100–125 VENT, SUB 15–25 and anal divided. **DISTRIBUTION** Coastal WA, from Cervantes north to Onslow. **HABITAT AND HABITS** Lives on sandy soils. Shelters beneath cover and leaf litter in coastal dune assemblages, heaths and sand-ridge deserts. Nocturnal. Feeds on ground-dwelling skinks. Lays 3–4 eggs. HARMFUL

Dampierland Burrowing Snake ■ *Simoselaps minimus* TL 41cm

DESCRIPTION Top of body pale yellowish-brown. Edges of midbody scales darker, forming reticulated pattern. Head has black band over nostrils and eyes, as well as nuchal band. Head and neck background colouration whitish to cream. Belly creamish-white. MB 15 rows, 125–135 VENT, SUB 19–25 and anal divided. **DISTRIBUTION** Northern WA on Dampier Peninsula north of Broome. **HABITAT AND HABITS** Lives on sandy soils. Shelters beneath cover and leaf litter in coastal dune assemblages, heaths and sand-ridge deserts. Nocturnal. Probably feeds on ground-dwelling skinks. Likely to lay eggs. HARMFUL

Rosen's Snake ■ *Suta fasciata* TL 45cm

DESCRIPTION Top of body reddish-brown to orange or cream, with maroon, brown or grey irregular markings and spotting. Belly creamish-white. MB 17–19 rows, 140–165 VENT, SUB 20–40 and anal single. Sometimes confused with Western Stimson's Python (see p. 19), differing by lacking labial pits. **DISTRIBUTION** Western and central WA, from Karratha to Kalgoorlie. **HABITAT AND HABITS** Lives on heavy clay and sandy soils in mulga, rocky gorges, open woodland and deserts. Found under rocks and fallen timber, and beneath leaf litter. Nocturnal. Feeds mainly on lizards, but will also eat small mammals and frogs. Produces 1–7 young. VENOMOUS

Ord Curl Snake ■ *Suta ordensis* TL 30cm

DESCRIPTION Top of body yellow-brown to dark brown or grey. Some individuals have dark edges to posterior of scales, giving reticulated appearance. Head usually darker than

rest of body, but this fades with maturity. Some individuals have light barring on lips. Belly white to cream. MB 19 rows, 165–185 VENT, SUB 30–40 and anal single. Lacks dark lateral head-streak from nostril to temporal region found in the Curl Snake (see opposite). **DISTRIBUTION** WA, in Ord and Victoria River catchments. **HABITAT AND HABITS** Lives in tropical woodland and black-soil grassland. Nocturnal. Ecology thought to be similar to that of the Curl Snake. VENOMOUS

Little Spotted Snake ■ *Suta punctata* TL 60cm

DESCRIPTION Top of body yellow-brown, orange to dark brown. Head, nape and forebody have a series of dark spots and blotches that usually fade with maturity. Usually dark brown stripe running along lower flanks. Some individuals have dark edges to

posterior of scales, giving reticulated appearance. Belly white to cream. MB 15 rows, 150–215 VENT, SUB 20–40 and anal single. **DISTRIBUTION** WA north of Quobba, through most of NT, to north-west Qld. **HABITAT AND HABITS** Lives in open woodland, gorges and escarpments, mulga and deserts. Seen actively hunting or crossing roads at night. Nocturnal. Diet comprises lizards, frogs and small mammals. Produces 2–5 young. VENOMOUS

Curl Snake ▪ *Suta suta* TL 80cm
(Myall Snake)

DESCRIPTION Top of body yellow-brown to dark brown or grey. Some individuals have dark edges to posterior of scales, giving reticulated appearance. Head usually darker than rest of body, but this fades with maturity. Some individuals have light barring on lips and broken yellow stripe along side of head. Dark lateral head-streak from nostril to temporal region. Belly white to cream. MB 19–21 rows, 150–170 VENT, SUB 20–35 and anal single. **DISTRIBUTION** Most of arid Qld, NSW, northern Vic, most of SA and NT. Isolated population near Lake Argyle, WA. **HABITAT AND HABITS** Lives in dry forests, mallee, heaths and deserts. Seen actively hunting or crossing roads at night. Found under logs and rocks, and beneath man-made debris, or in deep soil cracks; also in trees, hunting small dragons. Nocturnal. Diet comprises lizards, frogs and small mammals. Produces 1–9 live young. Polyvalent Antivenom is used to neutralize bites from this species. DANGEROUSLY VENOMOUS

Rough-scaled Snake ■ *Tropidechis carinatus* 90cm
(Clarence River Snake)

DESCRIPTION Top of body brown to grey, occasionally with greenish wash. Some individuals flecked with black. Northern animals can be completely banded. Belly yellowish-green. MB 23 rows, 160–180 VENT, SUB 50–60 and anal single.
DISTRIBUTION Two populations. Southern population occurs from Gosford, NSW, to Fraser Island, Qld. Northern population inhabits wet tropics region of north Qld.
HABITAT AND HABITS Lives in closed forests, wallum swamps and wetlands. Often found sitting in vegetation or crossing roads at night. Seen sheltering under cover, for

example beneath logs, in tree hollows, and under rocks or man-made objects. Nocturnal, occasionally basking in the morning or after rain. Diet comprises small mammals, frogs, tadpoles, small lizards and occasionally birds. Produces 5–19 young. Very toxic venom that has caused death extremely rapidly. Tiger Snake or Polyvalent Antivenom is used to neutralize bites from this species.

Northern population   DANGEROUSLY VENOMOUS

Bandy-bandy ■ *Vermicella annulata* TL 70cm

DESCRIPTION Top and underside of body a series of alternating black-and-white rings; 36–38 white rings. White ring count, when used with distribution, distinguishes each bandy-bandy species from others. MB 15 rows, 216–220 VENT, SUB 12–30 and anal divided. **DISTRIBUTION** Eastern Australia from Cape York, Qld, south to northern Vic, and west to Port Augusta, SA. **HABITAT**

AND HABITS Lives in forests, open woodland, mallee, brigalow, mulga and rocky areas in arid zones. Found under rocks and fallen timber, and beneath leaf litter. Nocturnal. Feeds on blind snakes and elongated skinks. Lays 4–6 eggs. Unusual many-banded population around Townsville has a higher white band count. Ingenious defensive strategy involving raising parts of body to form loops, which along with thrashing is employed to deter predators. Bites can cause pain, swelling and usually minor systemic effects such as headache and nausea. HARMFUL

Intermediate Bandy-bandy ■ *Vermicella intermedia* TL 39cm

DESCRIPTION Top and underside of body a series of alternating black-and-white rings; 50–53 white rings. White ring count, when used with distribution, distinguishes each bandy-bandy species from others. MB 15 rows, 246–256 VENT, SUB 15–28 and anal divided. **DISTRIBUTION** From Top End, NT, west across into Kimberleys, northern WA.

HABITAT AND HABITS Lives in tropical savannah, vine thickets and rocky areas. Found under rocks and fallen timber, and beneath leaf litter. Nocturnal. Probably feeds on blind snakes. Likely to lay eggs. Uses the same defensive strategy as the Bandy-bandy (see above). HARMFUL

Narrow-banded Bandy-bandy ■ *Vermicella multifasciata* TL 45cm

DESCRIPTION Top of body black to dark brown, with white rings formed by white spots arranged in band; 77–109 white rings. White ring count, when used with distribution,

distinguishes each bandy-bandy species from others. MB 15 rows, 240–296 VENT, SUB 15–25 and anal divided. **DISTRIBUTION** From Ord River drainage across into western Top End, NT, and on Melville and Bathurst Islands. **HABITAT AND HABITS** Nocturnal. Lives in tropical savannah. Found under rocks and fallen timber, and beneath leaf litter. Probably feeds on blind snakes. Likely to lay eggs. Uses the same defensive strategy as the Bandy-bandy (see p. 131). HARMFUL

Cape York Bandy-bandy ■ *Vermicella parscauda* TL 39cm

DESCRIPTION Top and underside of body a series of alternating black-and-white rings; 51–89 white rings. White ring count, when used with distribution, distinguishes each

bandy-bandy species from others. MB 15 rows, 213–230 VENT, SUB 27 and anal divided. **DISTRIBUTION** Cape York, between Weipa and Malpoon, Qld. **HABITAT AND HABITS** Lives in tropical open woodland on heavy red soils. Poorly known, and only described in 2018. Nocturnal. Probably feeds on blind snakes. Likely to lay eggs. HARMFUL

Pilbara Bandy-bandy ■ *Vermicella snelli* TL 41cm

DESCRIPTION Top and underside of body a series of alternating black-and-white rings; 48–64 white rings. White ring count, when used with distribution, distinguishes each bandy-bandy species from others. MB 15 rows, 262–302 VENT, SUB 12–30 and anal divided. **DISTRIBUTION** Pilbara region of WA. **HABITAT AND HABITS** Lives in gorges and rocky areas in arid zones. Found under rocks and beneath leaf litter. Nocturnal. Feeds on blind snakes. Likely to lay eggs. Uses the same defensive strategy as the Bandy-bandy (see p. 131). HARMFUL

Centralian Bandy-bandy ■ *Vermicella vermiformis* TL 32cm

DESCRIPTION Top and underside of body a series of alternating black-and-white rings; 41–45 white rings. White ring count, when used with distribution, distinguishes each bandy-bandy species from others. MB 15 rows, 263–281 VENT, SUB 12–30 and anal

divided. **DISTRIBUTION** Central Australia surrounding Alice Springs, NT; also in southern gulf country across to western Qld. **HABITAT AND HABITS** Lives in open woodland, gorges and rocky areas in arid zones. Found under rocks and fallen timber, and beneath leaf litter. Nocturnal. Feeds on blind snakes. Lays 9 eggs. Uses the same defensive strategy as the Bandy-bandy (see p. 131). HARMFUL

Homalopsidae (Mangrove Snakes)
For many years treated as a subfamily of the Colubridae, mangrove snakes are a group of mildly venomous, rear-fanged snakes that are either fully or partly aquatic. They are much more diverse in Southeast Asia than in Australia, with five of the approximately 54 species in the family occurring in northern Australia. Most species live in mangroves, emerging at night to hunt in intertidal zones and pools.

Australian Bockadam ▪ *Cerberus australis* TL 110cm

DESCRIPTION Two main colour phases, one reddish-orange, the other grey with variable amounts of black cross-bands and blotches. Usually a black temporal streak. Whitish-cream with darker spotting below. Dorsal scales keeled except on head. Eyes on top of head, and nostrils with valves, are adaptations to an aquatic environment. MB 23–25 rows, 140–160 VENT, SUB 45–60 all divided and anal divided. **DISTRIBUTION** Across northern and eastern Australia from Kalumburu, WA, to western coast of Cape York Peninsula, Qld. Also PNG. **HABITAT AND HABITS** Semi-aquatic, living in mangroves, along waterways in brackish and salt water. Predominantly nocturnal. Lives among mangrove roots and fallen timber, emerging on to the flats to hunt fish in small pools. Gives birth to 6–9 young. Rear fanged, and bites can result in stinging and minor swelling. HARMFUL

White-bellied Mangrove Snake ■ *Fordonia leucobalia* TL 90cm

DESCRIPTION Very variable; colouration a combination of whites, yellows, browns, reds, blacks, greys and orange. Can be patterned with dots, stripes or blotches, or plain. Lower flanks and underside usually white, and white with dark medial streak beneath tail. No loreal scale. MB 23–29 rows, 130–160 VENT, SUB 25–45 all divided and anal divided.

DISTRIBUTION Across northern and eastern Australia from Nichol Bay, WA, to western coast of Cape York Peninsula and Torres Strait Islands, Qld. Also PNG and Southeast Asia.

HABITAT AND HABITS Semi-aquatic, living in mangroves, along waterways in brackish and salt water. Lives among mangrove roots and fallen timber, sheltering in crab holes. Predominantly nocturnal. Emerges on to the flats to hunt crabs and small lobsters. One of two Australian snakes that dismember their prey, removing the legs before eating the body. Gives birth to 2–17 young. Rear fanged, and bites can result in stinging. HARMFUL

Roebuck Bay Mangrove Snake ■ *Myron resetari* TL 40cm

DESCRIPTION Silver-grey, brown to olive, with series of black cross-bands and blotches. Usually a black temporal streak. Whitish-cream to yellow below. Dorsal scales keeled except on head. MB 19 rows, 137–145 VENT, SUB 30–40 all divided and anal divided.

Differs from the similar Richardson's Mangrove Snake (see below) by only having single preocular scale on each side of head vs 2, and fewer MB rows. **DISTRIBUTION** Currently only around Roebuck Bay and Broome, WA. Expected to occur over large expanse of WA coastline. **HABITAT AND HABITS** Semi-aquatic, living in mangroves, along waterways in brackish and salt water. Predominantly nocturnal. Lives among mangrove roots and in burrows, emerging on to the flats to hunt gobies in small pools. Recently described, in 2011. Thought to give birth to live young. Rear fanged, and bites unknown. HARMFUL

Richardson's Mangrove Snake ■ *Myron richardsonii* TL 50cm

DESCRIPTION Silver-grey, brown to olive-green, with irregular black cross-bands and blotches. Usually a black temporal streak. Some individuals in Torres Strait reddish-orange or black. Whitish-cream or yellow below. Dorsal scales keeled except on head. MB 21–23 rows, 130–147 VENT, SUB 30–40 all divided and anal divided. Differs from the similar Roebuck Bay Mangrove Snake (see above) by having 2 preocular scales on each side of head vs 1, and more MB rows.

DISTRIBUTION Across northern and eastern Australia from Derby, WA, to western coast of Cape York Peninsula and Torres Strait Islands, Qld. Also PNG and Indonesia. **HABITAT AND HABITS** Semi-aquatic, living in mangroves, along waterways in brackish and salt water. Predominantly nocturnal. Lives among mangrove roots and in burrows, emerging on to the flats to hunt gobies in small pools and shallow water. May also feed on molluscs. Gives birth to 6–9 young. Rear fanged, and bites can result in minor stinging. HARMFUL

Macleay's Water Snake ■ *Pseudoferania polylepis* TL 110cm

DESCRIPTION Dark grey to pale brown. Usually striped with darker tones. Lower flanks cream to yellow. Often flecked with dark speckling. Underside cream to bright yellow, with heavy dark flecking. MB 21–23 rows, 140–165 VENT, SUB 35–50 all divided and anal divided. **DISTRIBUTION** Two separate populations. One extends through northern NT across to Riversleigh, Qld. The other occurs from Georgetown across to Ingham, Qld, north across Torres Strait. Also PNG. **HABITAT AND HABITS** Aquatic, living

Western population

in freshwater rivers, swamps and billabongs. Rarely found on land. Predominantly nocturnal. Lives among tree roots and vegetation, where it is an ambush predator of fish including eels, and occasionally frogs. Gives birth to 4–27 young. Possibly two species. Rear fanged, and bites can result in stinging and minor swelling. HARMFUL

Eastern population

Small-headed Blind Snake ■ *Anilios affinis* TL 23cm

DESCRIPTION Pale brown to pinkish above, and lighter on head. White or cream to yellowish below. MB 18 rows, 278–357 DSR, SUB 10–19. Body type slender with head rounded when viewed from above and angular in profile. **DISTRIBUTION** Eastern Qld,

from Charters Towers to just over NSW border. Also Solomon Islands, but these snakes probably represent a morphologically similar but different species. **HABITAT AND HABITS** Fossorial, living in open woodland on clay and loamy soils. Known to shelter in termite mounds. Nocturnal. Recorded as laying 3 eggs. HARMLESS

Sand-diving Blind Snake ■ *Anilios ammodytes* TL 35cm

DESCRIPTION Dark purplish-brown to pinkish-brown above, and darker on head. Lighter in colour below. MB 20 rows, 389–498 VENT, SUB 8–18. Body type slender with head rounded when viewed

from above and in profile. **DISTRIBUTION** Pilbara region, WA, from North West Cape to Eighty Mile Beach. **HABITAT AND HABITS** Fossorial, living in open woodland, grassland and coastal dunes with spinifex and mulga, on both stony and sandy soil types. Nocturnal. Known to shelter beneath logs and rocks. HARMLESS

No-spined Blind Snake ■ *Anilios aspina* TL 28cm
(Round-tailed Blind Snake)

DESCRIPTION Brown to pink in colouration. Posterior edges of midbody scales can be darker. Lighter below. MB 18 rows, 437 VENT, SUB 14. Body type slender with head rounded when viewed from above and in profile. One of two Australian blind snakes that lacks a terminal (apical) spine (see West Kimberley Blind Snake, p. 140). **DISTRIBUTION** Central Qld, from Julia Creek to Barcaldine. **HABITAT AND HABITS** Fossorial, living in Mitchell grassland on heavy clay soils and black-soil plains. Nocturnal. Poorly known from a few individuals. HARMLESS

Southern Blind Snake ■ *Anilios australis* TL 46cm

DESCRIPTION Dark purplish-brown to pinkish-brown above, and lighter on head. Paler below. MB 22 rows, 278–357 VENT, SUB 10–18. Body type slender with head rounded when viewed from above and angular in profile. Northernmost individuals often have pale collar on nape, and may require further examination to determine status. **DISTRIBUTION** South-west WA, from Cape Arid to Shark Bay. **HABITAT AND HABITS** Fossorial, living in open woodland, mallee, heaths, grassland, and coastal dunes with spinifex and mulga. Predominantly on sandy soil types. Nocturnal. Known to shelter beneath stumps and rocks, and in stick-ant nests. Lays 2–11 eggs. HARMLESS

Shovel-snouted Blind Snake ■ *Anilios batillus* TL 32cm

DESCRIPTION Yellowish-brown with evidence of dark pigment forming dark longitudinal stripes. Underside paler, being yellow to cream. MB 24 rows, 557 DSR, SUB 21. Body type slender with head bluntly pointed when viewed from above and angular in profile. **DISTRIBUTION** Known from single individual from Wagga Wagga, NSW, described in 1894. **HABITAT AND HABITS** Fossorial, living in open woodland. Presumed to be nocturnal. The collection location is possibly an error and it may not be an Australian species. HARMLESS

Dark-spined Blind Snake ■ *Anilios bicolor* TL 42cm

DESCRIPTION Dark purplish-brown to brown above. Lightens on lower flanks, merging to white underneath. MB 22 rows, 357 VENT, SUB 14. Body type robust with head weakly

trilobed when viewed from above and angular in profile. **DISTRIBUTION** Southern WA, from Cocklebiddy, across southern Australia, to Lake Cargelligo, NSW; also north-west Vic. **HABITAT AND HABITS** Fossorial, living in open woodland and mallee with spinifex and mulga. Predominantly on sandy soil types. Nocturnal. Known to shelter under stumps and rocks, and beneath accumulated leaf litter. Can make an audible squeak if grasped. Lays 6 eggs. HARMLESS

Prong-snouted Blind Snake ■ *Anilios bituberculatus* TL 46cm

DESCRIPTION Dark purplish-black to pink above. Paler below. MB 20 rows, 414–485 VENT, SUB 11–18. Body type slender with head strongly trilobed when viewed from

above and angular in profile. **DISTRIBUTION** Across arid and semi-arid southern Australia, from WA coastline, through SA, into western Vic, NSW and southern Qld. **HABITAT AND HABITS** Fossorial, living in open woodland, mallee, heaths, and coastal dunes with spinifex and mulga. Predominantly on sandy soil types. Nocturnal. Known to shelter beneath stumps, rocks and leaf litter, and inside termite nests. Lays 3–7 eggs. HARMLESS

Faint-striped Blind Snake ■ *Anilios broomi* TL 26cm

DESCRIPTION Dark pinkish-brown above with darker centres aligning to thin dorsal stripes running along length of body. Usually darker on head. White below. MB 20 rows, 456–460 VENT, SUB 15–16. Body type slender with head rounded when viewed from above and in profile. **DISTRIBUTION** From Cooktown to Innot Hot Springs along Einasleigh Uplands, Qld. **HABITAT AND HABITS** Fossorial, living in open woodland. Nocturnal. Known to shelter beneath rocks and fallen timber. HARMLESS

Centralian Blind Snake ■ *Anilios centralis* TL 30cm

DESCRIPTION Dark pinkish-brown to yellowish-brown above. Paler below. MB 20 rows, 417–502 VENT, SUB 12–20. Body type slender with head rounded when viewed from above, and pointed with weakly recurved, hook-shaped rostral in profile. **DISTRIBUTION** Central Australia around Alice Springs and MacDonnell Ranges, NT. **HABITAT AND HABITS** Fossorial, living in rocky gorges with spinifex and low woodland. Predominantly on sandy soil types. Nocturnal. Lays 5 eggs. HARMLESS

Cape York Striped Blind Snake ■ *Anilios chamodracaena* TL 21cm

DESCRIPTION Dark pinkish-brown above with darker centres aligning to thin dorsal stripes running along length of body. Usually darker on head. White below. MB 18 rows,

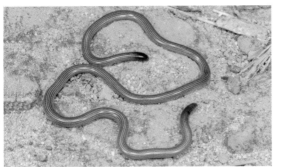

464–523 VENT, SUB 14–16. Body type slender with head rounded when viewed from above and in profile. **DISTRIBUTION** From Weipa to Inkerman Station on Cape York Peninsula, Qld. **HABITAT AND HABITS** Fossorial, living in open woodland. Nocturnal. Known to shelter beneath rocks and inside rotten timber. HARMLESS

Northern Blind Snake ■ *Anilios diversus* TL 36cm

DESCRIPTION Dark purplish-brown to pinkish-brown above. Paler below. MB 20 rows, 384–457 VENT, SUB 8–18. Body type slender with head rounded when viewed from above and in profile. **DISTRIBUTION** From Morven, Qld, through NT, to Kimberley region, WA. **HABITAT AND HABITS** Fossorial, living in open woodland, black-soil grassland and coastal savannah. Predominantly on sandy soil types. Nocturnal. Known to shelter beneath rocks and fallen timber. Lays 5–8 eggs. HARMLESS

Interior Blind Snake ■ *Anilios endoterus* TL 36cm

DESCRIPTION Dark purplish-brown to pinkish-brown above, sometimes reddish. Cream to white below. MB 22 rows, 406–438 VENT, SUB 9–16. Body type slender with head moderately trilobed when viewed from above and gently angular in profile. **DISTRIBUTION** Central Australia from top of Eyre Peninsula, SA, through western NSW and Qld, southern NT, and much of WA interior. **HABITAT AND HABITS** Fossorial, living in open woodland, mallee, grassland and deserts. Predominantly on sandy soil types. Nocturnal. Known to shelter beneath fallen timber and rocks. HARMLESS

Miner Blind Snake ■ *Anilios fossor* TL 25cm

DESCRIPTION Pale yellowish-brown above. Paler below. Colouration in life may be pink. MB 20 rows, 514 DSR, SUB 11. Body type slender with head rounded when viewed from above and rounded in profile. **DISTRIBUTION** Known from a couple of individuals found at Glen Annie in Ruby Gap Nature Reserve, southern NT, about 100km east of Alice Springs. **HABITAT AND HABITS** Fossorial, living in open river gum woodland with *Acacia* on sandy soil types. Nocturnal. HARMLESS

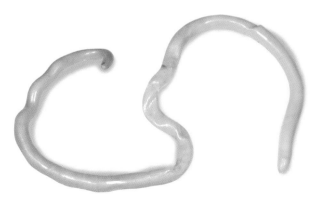

Gane's Blind Snake ■ *Anilios ganei* TL 34cm

DESCRIPTION Dark greyish-brown to brown above. White to cream below. MB 24 rows, 406–448 VENT, SUB 12–19. Body type moderately robust with head rounded when viewed from above and in profile. **DISTRIBUTION** WA in Pilbara region between Newman, Millstream and Pannawonica. **HABITAT AND HABITS** Fossorial, living in moist shady gorges. Nocturnal. Noted in townships and while crossing roads. HARMLESS

Long-beaked Blind Snake ■ *Anilios grypus* TL 42cm

DESCRIPTION Pale brown to brown above with black tail. Snout pale in most populations, and black head. Paler below. MB 18 rows, 526–677 VENT, SUB 13–36. Body type very slender with head angular when viewed from above and deeply hooked

in profile. **DISTRIBUTION** WA, from Shark Bay to Kimberley, across Australia into central Qld. **HABITAT AND HABITS** Fossorial, living in open woodland, grassland and coastal dunes with spinifex and mulga. Nocturnal. Known to shelter under rocks, in termite nests and fallen timber, and beneath leaf litter. Genetics suggest that more than one species is involved. HARMLESS

Top End Blind Snake ■ *Anilios guentheri* TL 30cm

DESCRIPTION Dark purplish-brown to pinkish-brown above. Lighter on head. Paler below. MB 18 rows, 564–610 VENT (western), 464–547 (eastern), SUB 10–15. Body type slender with head rounded when viewed from above and in profile. **DISTRIBUTION** WA, from southern Kimberley across into north-western part of NT. **HABITAT AND HABITS** Fossorial, living in tropical open woodland, grassland and escarpment margins. Predominantly on sandy soil types. Nocturnal. Known to shelter beneath fallen timber and under leaf litter. Lays 2–4 eggs. Eastern and western populations may prove to be different taxa, with consistent differences in ventral counts. More research is required. HARMLESS

Pale-headed Blind Snake ■ *Anilios hamatus* TL 37cm

DESCRIPTION Brown to pinkish-brown above. Rostral lighter in colour. Paler below. MB 22 rows, 330–396 VENT, SUB 11–22. Body type moderately slender with head weakly trilobed when viewed from above and angular in profile, terminating in hooked rostral scale. **DISTRIBUTION** Central western WA, from Hamersley Range south to Merredin, and to Shark Bay and east across to Kalgoorlie region. **HABITAT AND HABITS** Fossorial, living in open woodland, mallee, heaths, and saltbush with spinifex and mulga. Predominantly on compacting loam and stony soil types. Nocturnal. Known to shelter beneath rocks and in stick-ant nests. HARMLESS

Kimberley Deep-soil Blind Snake ■ *Anilios howi* TL 23cm

DESCRIPTION Dark purplish-brown to blackish-brown above. Paler brown below. MB 18 rows, 430–533 VENT, SUB 10–16. Body type slender with head rounded when viewed from above and in profile. **DISTRIBUTION** Kimberley region, WA, from Kalumburu to Walsh Point in Admiralty Gulf. **HABITAT AND HABITS** Fossorial, living in tropical woodland and coastal plains. Found on damp reddish clay and stony soils. Nocturnal. HARMLESS

Fassifern Blind Snake ■ *Anilios insperatus* TL 9.6cm

DESCRIPTION Pale yellow in colour in preservative – the snake was light pink in life, due to oxygenation of the blood, without evidence of pigmentation on the body. MB 16 rows, 442 DSR and SUB 19. Body type slender with head weakly trilobed when viewed from

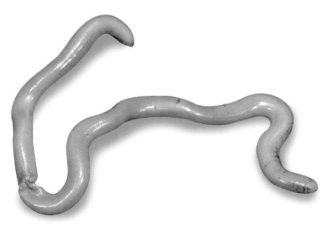

above and bluntly angular in profile. **DISTRIBUTION** South-east Qld at Warrill View. **HABITAT AND HABITS** Known from a single individual, found beneath a small stone on a cleared hillside that was previously open woodland of Lemon-scented Gum, Red Ironbark and Grey Box. Soil type is shallow, hard-setting clay. Currently Australia's smallest known snake. HARMLESS

Kimberley Shallow-soil Blind Snake ■ *Anilios kimberleyensis* TL 30cm

DESCRIPTION Dark brown to grey-brown above, merging from lower flanks to creamish below. Head slightly flattened and lighter in colour. MB 22 rows, 405–504 VENT, SUB 10–16. Body type slender with head rounded when viewed from above and in profile. **DISTRIBUTION** Kimberley region, WA, from Koolan Island to Kalumburu, and north-western NT from border across to Litchfield NP. **HABITAT AND HABITS** Fossorial, living in tropical woodland and moist vine thickets in gorges. Nocturnal. HARMLESS

Murchison Blind Snake ■ *Anilios leptosoma* TL 27cm

DESCRIPTION Dark purplish-brown to pinkish-brown above. Lighter on head. Paler below. MB 18 rows, 583–781 VENT, SUB 30 with 4 anal scales. Body type slender with head weakly trilobed when viewed from above and bluntly angular in profile, the rostral forming a weak hook. **DISTRIBUTION** WA, from Northampton to Wooramel, in Murchison region. **HABITAT AND HABITS** Fossorial, living in open woodland, mallee, heaths and coastal dunes. Predominantly on sandy soil types. Nocturnal. Known to shelter beneath fallen timber and rocks. HARMLESS

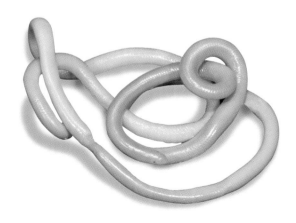

Cape York Blind Snake ■ *Anilios leucoproctus* TL 25cm

DESCRIPTION Dark purplish-black above, gradually becoming pale on lower flanks, to brown below. Some are dark purplish-black all over. MB 20 rows, 386–426 DSR, 14–17

SUB. Body type slender with head rounded when viewed from above and in profile. Similar to the Flowerpot Snake (see p. 163), which differs by having a smaller adult size and light-coloured ventral surface. **DISTRIBUTION** Qld, from Mt Tozer to tip of Cape York and some islands of Torres Strait. Also southern PNG. **HABITAT AND HABITS** Fossorial, living in lowland forests often with termite colonies. Predominantly on sandy soil types. Nocturnal. Known to shelter beneath stumps and rocks, and in ant nests. HARMLESS

Robust Blind Snake ■ *Anilios ligatus* TL 57cm

DESCRIPTION Dark purplish-brown to blackish-brown above. Some individuals slightly lighter on head. Hatchlings pinkish-purple. Yellowish to white below. MB 24 rows, 296–355 VENT, SUB 11–17. Body type very robust with head rounded when viewed from above and in profile. Ventral counts in western population higher than in eastern population. **DISTRIBUTION** Two separated populations. Western population extends from Kimberley region, WA, across NT, and into north-western Qld to about Mt Isa. Eastern population extends through eastern Qld, from Weipa into northern NSW, to about Gunnedah. **HABITAT AND HABITS** Fossorial, living in closed and open woodland, grassland and brigalow. Predominantly on sandy soil types. Nocturnal. Has been encountered diurnally active on the surface after rain. Known to shelter beneath fallen timber and rocks, and in ant nests. Lays 2–13 eggs that take 37 days to hatch. HARMLESS

Western population   *Eastern population*

Barrow Island Blind Snake ■ *Anilios longissimus* TL 27cm

DESCRIPTION Pale yellow in colour in preservative – the specimen was semi-translucent and appeared light pink in life due to oxygenation of the blood. MB 16 rows, 750 vertebral scales, SUB 15. Body type very slender with head strongly depressed, squarish when viewed from above and rounded in profile. **DISTRIBUTION** Bandicoot Bay, Barrow Island, WA.

HABITAT AND HABITS Potentially a true troglodyte living in reddish soil in subterranean limestone caverns. Habits largely unknown. Two individuals known, one escaping below the surface, while the other was collected and sent to the Western Australian Museum. It was found on the outer casing of a bore, removed from deep below the surface. A similar snake was found in Exmouth and was thought to be a second individual, but is actually an undescribed taxon. HARMLESS

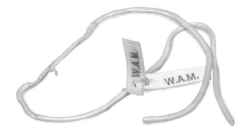

Buff-snouted Blind Snake ■ *Anilios margaretae* TL 30cm

DESCRIPTION Dark purplish-brown to pinkish-brown above, lighter on head. Pale grey below. MB 18 rows, 559 VENT, SUB 12. Body type very slender with head weakly trilobed when viewed from above and angular in profile. **DISTRIBUTION** Southern central WA at Lake Throssell, and near Maralinga, SA. **HABITAT AND HABITS** Fossorial, living at salt-lake margins with reddish sandy loams. Nocturnal. HARMLESS

Small-eyed Blind Snake ■ *Anilios micromma* TL 21cm

DESCRIPTION Pale yellow in colour in preservative – may have appeared light pink in life due to oxygenation of the blood. MB 18 rows, 478 VENT, SUB 15. Body type slender with head rounded when viewed from above and in profile. Eyes very small.

DISTRIBUTION Southern Kimberley region, WA, from Leopold Downs Station. **HABITAT AND HABITS** Largely unknown. Currently only known from a single individual collected in 1924. Habitat on the station is typical tropical open woodland. It has been suggested that the eyes are normal in size and that the lack of pigment surrounding them has led to the assumption of a smaller eye size. HARMLESS

Groote Eylandt Dwarf Blind Snake ■ *Anilios minimus* TL 22cm

DESCRIPTION Dark pinkish-brown with dark, longitudinal streaks. Head and tail dark in colour. Pale pinkish-white below. MB 16 rows, 381–457 VENT, SUB 9–17. Body

type slender with head rounded when viewed from above and in profile. **DISTRIBUTION** Groote Eylandt and adjacent areas of NT, to Lake Evella. **HABITAT AND HABITS** Fossorial, living in tropical woodland and coastal plains. Nocturnal. Found on sandy soils; one individual came from a coastal dune, beneath an embedded log on moist sandy soil. HARMLESS

Thread-like Blind Snake ■ *Anilios nema* TL 28cm
(Slender Blind Snake)

DESCRIPTION Dark purplish-brown to pale pink. Pale pinkish-white below. MB 16
rows, 520–589 VENT, SUB
9–14. Body type slender
with head rounded when
viewed from above and in
profile. **DISTRIBUTION**
In and around Darwin,
NT. **HABITAT AND
HABITS** Fossorial, living
in tropical woodland and
heavily vegetated gardens.
Nocturnal. Found beneath
rocks and fallen timber,
and in household building
materials. HARMLESS

Blackish Blind Snake ■ *Anilios nigrescens* TL 75cm

DESCRIPTION Black, dark purplish-brown to pinkish-brown above. Juveniles pink. Edge
of scales whitish, forming a fine netted pattern. Underside white. MB 22 rows, 420–457
VENT, SUB 13–19. Body type moderately robust with head rounded when viewed
from above and in profile. **DISTRIBUTION** Southern Qld, from Kroombit Tops, south
through eastern NSW, to Seymour, Vic. **HABITAT AND HABITS** Fossorial, living in
open and closed woodland, rocky outcrops, heaths, grassland and rainforests. Nocturnal.

Predominantly on
sandy soil types.
Known to shelter
beneath stumps and
rocks, and among
leaf litter, sometimes
forming aggregations
of multiple individuals.
Frequently encountered
in suburban gardens
and swimming pools,
or crossing roads on
warm, wet nights.
Anecdotal reports of
it eating worms and
leeches. Lays 5–20 eggs.
HARMLESS

Blunt-snouted Blind Snake ■ *Anilios obtusifrons* TL 22.5cm

DESCRIPTION Pinkish in colour, becoming lighter on lower flanks. Underside paler, being yellow to cream. MB 18 rows, 581–590 VENT, SUB 15 and anal 4. Body type

slender with head bluntly pointed when viewed from above and angular in profile. **DISTRIBUTION** Known from three individuals found in two locations south of Kalbarri, WA. **HABITAT AND HABITS** Fossorial, living in *Acacia* woodland, heaths and mallee with brown sandy soil. Presumed to be nocturnal. HARMLESS

Pilbara Blind Snake ■ *Anilios pilbarensis* TL 38cm

DESCRIPTION Brown to purplish-brown above or grey gradually merging into whitish ventral surface. Rostral lighter in colour. Paler below. MB 22 rows, 363–425 VENT, SUB 15–22. Body type moderately slender with head weakly trilobed when viewed from above and strongly angular in profile. **DISTRIBUTION** Central western WA, from Muccan

to Balfour Downs across to Chichester Range NP. **HABITAT AND HABITS** Fossorial, living in open woodland with spinifex. Also found in areas with *Acacia* with granite outcrops. Predominantly on compacting loam soil types. Nocturnal. Known to shelter beneath rocks and fallen timber. HARMLESS

Rotund Blind Snake ■ *Anilios pinguis* TL 55cm

DESCRIPTION Dark purplish-brown to black above, lighter on head. Lower flanks become gradually lighter, turning pale grey below. MB 20 rows, 280–377 VENT, SUB 10–18. Body type very robust with head weakly trilobed when viewed from above and bluntly angular in profile.

DISTRIBUTION Southwest WA, from Gingin to Bruce Rock across to Bunbury. **HABITAT AND HABITS** Fossorial, living in open woodland, mallee, heaths and rock outcrops. Predominantly on heavily stony soil types. Nocturnal. Known to shelter beneath stumps and rocks, and in stick-ant nests. Lays about 5 eggs. HARMLESS

Proximus Blind Snake ■ *Anilios proximus* TL 75cm

DESCRIPTION Dark purplish-brown to brown above. Hatchlings pinkish-purple. Yellowish to white below. MB 22 rows, DSR 326–378, SUB 12–23. Body type very robust with head slightly trilobed when viewed from above and angular in profile. Atherton population has a higher DSR count (390–392 vs 326–378) than main population. **DISTRIBUTION** Two separated populations. Northern population centres around Atherton Tablelands, Qld. Main population extends through eastern Qld, from Rockhampton, through NSW and Vic to Macedon, and across to SA border.

HABITAT AND HABITS Fossorial, living in both closed and open woodland, grassland and brigalow. Predominantly on sandy soil types. Nocturnal, but has been encountered diurnally active on the surface after rain. Known to shelter beneath fallen timber and rocks, and in ant nests. Lays 5–13 eggs. HARMLESS

Roberts' Blind Snake ■ *Anilios robertsi* TL 29cm

DESCRIPTION Dark purplish-black above with sharp, straight-edged delineation to stark white underside. MB 22 rows, 556 VENT, SUB 12. Body type slender with head rounded

when viewed from above and in profile. Similar to the Flowerpot Snake (see p. 163), but this differs by not having sharp edge between dorsal and ventral colouration. **DISTRIBUTION** Known only from Romeo Creek near Shipton's Flat, north Qld. **HABITAT AND HABITS** Thought to be nocturnal. Known from single individual found in February 1983 coming out of a fallen hollow log in open woodland. HARMLESS

Cooloola Blind Snake ■ *Anilios silvia* TL 17.5cm

DESCRIPTION Broadly striped with dark purplish-black on yellowish-cream body. Stripes usually coalesce, giving it the appearance of a wholly dark purplish-black snake above with a sharp, variable-edged delineation to a stark white underside. MB 20 rows, 272–320 VENT, SUB 14–21. Body type slender with head rounded when viewed from above and in profile. **DISTRIBUTION** Found in south-east Queensland, from Palmwoods to Fraser Island, Qld. **HABITAT AND HABITS** Fossorial, living in closed woodland and rainforests. Predominantly on sandy soil types. Nocturnal. Known to shelter beneath fallen timber and rocks, and in rotten timber. HARMLESS

Splendid Blind Snake ▪ *Anilios splendidus* TL 51cm

DESCRIPTION Silver purplish-grey above with randomly placed white scales, lighter on front half of head. Paler below. MB 20 rows, 377 VENT, SUB 13. Body type moderately robust with head sub-rectangular and weakly trilobed when viewed from above and weakly angular in profile. **DISTRIBUTION** Ranger's residence (Milyering Well), Cape Range NP, WA. **HABITAT AND HABITS** Fossorial. Nocturnal. Known from single individual found active at night in April 1995. Habitat in the area was sand over coral limestone with Tangling Melaleuca, small shrubs and *Triodia*. HARMLESS

Sharp-snouted Blind Snake ■ *Anilios systenos* TL 27cm

DESCRIPTION Yellowish-brown with evidence of dark pigment forming dark longitudinal

stripes. Underside paler, being yellow to cream. MB 18 rows, 598–621 VENT, SUB 30 and 4 anal scales. Body type slender with head bluntly pointed when viewed from above and angular in profile. **DISTRIBUTION** Known from a few individuals just east of Geraldton, WA. **HABITAT AND HABITS** Fossorial, living in open woodland. Presumed to be nocturnal. HARMLESS

North-eastern Blind Snake ■ *Anilios torresianus* TL 40cm

DESCRIPTION Dark purplish-brown to pinkish-brown above, lighter on head. Paler below. MB 22 rows, 348–387 VENT, SUB 14–22. Body type robust with head rounded when viewed from above and in profile. **DISTRIBUTION** North-east Qld, from Mackay to tip of Cape York. Also southern PNG. **HABITAT AND HABITS** Fossorial, living in

open and closed woodland, vine thickets, rainforests and moist gardens. Nocturnal. Known to shelter beneath stumps and rocks, and in termite nests. Lays 2–7 eggs. Often known under the name *Ramphotyphlops polygrammicus* – this name is now restricted to snakes from Timor-Leste and Timor. HARMLESS

Darwin Blind Snake ■ *Anilios tovelli* TL 13cm

DESCRIPTION Dark brown to pinkish-brown above, with dark medial pigment on each scale, which can form faint stripes. Tail dark. Pinkish-grey below. MB 20 rows, 240–262

VENT, SUB 13–16. Body type slender with head rounded when viewed from above and in profile. Similar to the Thread-like Blind Snake (see p. 153), but differs by having higher midbody scale count (20 vs 16). **DISTRIBUTION** Darwin and surrounding areas, including Cobourg Peninsula, NT. **HABITAT AND HABITS** Fossorial, living in open woodland, vine thickets and floodplains. Nocturnal. Known to shelter beneath stumps and rocks. HARMLESS

Sandamara Blind Snake ■ *Anilios troglodytes* TL 40cm

DESCRIPTION Pale brown to pinkish above, lighter on head. White, cream to yellowish below. MB 22 rows, 587–641 VENT, SUB 12–14. Body type very slender with head

flattened and rounded when viewed from above and in profile. **DISTRIBUTION** North-western WA, from El Questro Station across to Tunnel Cave in Napier Range. **HABITAT AND HABITS** Fossorial, living in open woodland. Nocturnal. Known to shelter beneath stumps and rocks. HARMLESS

Claw-snouted Blind Snake ■ *Anilios unguirostris* TL 70cm

DESCRIPTION Dark purplish-brown to pinkish-brown above, darker on head. Lighter below. MB 24 rows, 387–474 VENT, SUB 11–16. Body type medium build with head

poorly trilobed when viewed from above, and strongly hook-like and angular in profile. **DISTRIBUTION** Kimberley, WA, across northern Australia, to south-central Qld. **HABITAT AND HABITS** Fossorial, living in open woodland, grassland and mulga land, on both stony and sandy soil types. Nocturnal. Known to shelter beneath logs and rocks. HARMLESS

Beaked Blind Snake ■ *Anilios waitii* TL 62cm

DESCRIPTION Pale yellowish-brown to dark brown above. Merges into pale greyish-cream below. Rostral usually pale. MB 20 rows, 535–667 VENT, SUB 13–26. Body type slender with head weakly trilobed when viewed from above and strongly angular in profile. **DISTRIBUTION** Southern WA, from Warburton Range, through goldfields and wheatbelt regions to Perth. **HABITAT AND HABITS** Fossorial, living in mallee, sand-ridge deserts and grassland, on red loams and pale sands. Nocturnal. Has been seen entering a meat-ant nest. HARMLESS

Brown-snouted Blind Snake ■ *Anilios weidii* TL 30cm

DESCRIPTION Dark purplish-brown to pinkish-brown above. Rostral scale usually brown. Paler below. MB 20 rows, 390–410 VENT, SUB 20. Body type slender with head rounded when viewed from above and in profile.
DISTRIBUTION Eastern Qld, from Mackay south into NSW to Hunter Valley, and through to plains of central NSW. **HABITAT AND HABITS** Fossorial, living in open woodland, brigalow, grassland and mulga. Predominantly on sandy soil types. Nocturnal. Known to shelter beneath stumps and rocks, and in leaf litter. Lays 1–8 eggs. HARMLESS

Yampi Blind Snake ■ *Anilios yampiensis* TL 13cm

DESCRIPTION Yellowish-brown with evidence of dark pigment on head and tail. Underside paler, being yellow to cream. MB 18 rows, 480 VENT, SUB 11. Body type slender with head bluntly pointed when viewed from above and angular in profile.
DISTRIBUTION Known from single individual found on Koolan Island off Yampi Peninsula in Kimberley region, WA, in March 1966.
HABITAT AND HABITS Presumed to be nocturnal and fossorial, and probably lives in gorges made up of quartzite sandstone with vine forests. HARMLESS

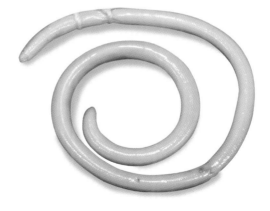

Yirrkala Blind Snake ■ *Anilios yirrikalae* TL 20cm

DESCRIPTION Dark purplish-brown to pinkish-brown above. Paler below. MB 24 rows, VENT 434, SUB 13. Body type slender with head rounded when viewed from above and in profile.

DISTRIBUTION Around Caledon Bay, Yirrkala Mission, in far north-east of Arnhem Land, NT. **HABITAT AND HABITS** Fossorial, living in open tropical woodland. Predominantly on sandy soil types. Nocturnal. Known to shelter beneath fallen timber and rocks. HARMLESS

West Kimberley Blind Snake ■ *Anilios zonula* TL 18cm

DESCRIPTION Dark purplish-brown to pinkish-brown above, slightly darker towards head. Paler below. MB 18 rows, 446–482 DSR, SUB 11–15. Body type slender with head rounded when viewed from above and in profile. Distinguished from all other local species by distribution and lack of apical spine (see No-spined Blind Snake, p. 140).

DISTRIBUTION Known from Augustus and Storr Islands in western Kimberley, WA, but likely to also occur on mainland close by. **HABITAT AND HABITS** Fossorial, living in vine thickets surrounded by open woodland. Nocturnal. Known to shelter beneath sandstone rock slabs. HARMLESS

Flowerpot Snake ■ *Indotyphlops braminus* TL 17cm

DESCRIPTION Brown to purplish-black. Lighter below. MB 18 rows, 261–358 DSR, VENT 300–331, SUB 8–15. Body type slender with head rounded when viewed from above and in profile. Conspicuous glands on head forming faint, scribbly pale lines.
DISTRIBUTION Introduced, arriving in Australia in the 1960s. Currently found in scattered locations across northern Australia, including Brisbane, Townsville, Torres Strait, Darwin, Derby, Pilbara coast, Perth and Christmas Island. Also throughout the tropics worldwide, possibly originating in Southeast Asia.

HABITAT AND HABITS Fossorial, living in gardens and forests. Populations in the Pacific are also arboreal. Nocturnal. Parthenogenetic, laying 2–11 eggs. Known as the Flowerpot Snake due to how it has been transported worldwide via plant movement. HARMLESS

Christmas Island Blind Snake ■ *Ramphotyphlops exocoeti* TL 35cm

DESCRIPTION Pink, with brown centres on each scale giving appearance of variegations or fine longitudinal stripes. Cream to white below. MB 20 rows, VENT 354, SUB 21. Body type slender with head rounded when viewed from above and in profile. **DISTRIBUTION** Found only on Christmas Island in tropical Indian Ocean.

HABITAT AND HABITS Fossorial, living in rainforests among limestone casts. Nocturnal. Most individuals found under fallen timber after rain. At risk of extinction due to inadvertent introduction of multiple exotic predators on the island. HARMLESS

CHECKLIST OF THE SNAKES OF AUSTRALIA

Taxonomy follows Cogger 2018, with the exception of the addition of newly described taxa. For each species, an 'x' indicates presence in a particular state or territory.

State or territory abbreviations:

NSW	New South Wales (including Australian Capital Territory)
NT	Northern Territory
Qld	Queensland
SA	South Australia
Tas	Tasmania
Vic	Victoria
WA	Western Australia

Abbreviations of IUCN Red List status:

EX	Extinct
CR	Critically Endangered
EN	Endangered
VU	Vulnerable
NT	Near Threatened
LC	Least Concern
DD	Data Deficient
NE	Not Evaluated
UK	Unknown

Common English Name	Scientific Name	Qld	NSW	Vic	Tas	SA	WA	NT	Territories	IUCN
Pythons (Pythonidae)										
Children's Pythons, Genus *Antaresia*										
Children's Python	Antaresia childreni	x					x	x		LC
Spotted Python	Antaresia maculosa	x	x							LC
Pygmy Python	Antaresia perthensis						x			LC
Western Stimson's Python	Antaresia stimsoni stimsoni						x			LC
Stimson's Python	Antaresia stimsoni orientalis	x	x			x	x	x		LC
Black headed and Woma Pythons, Genus *Aspidites*										
Black-headed Python	Aspidites melanocephalus	x					x	x		LC
Woma	Aspidites ramsayi	x	x			x	x	x		LC
Olive and Water Pythons, Genus *Liasis*										
Water Python	Liasis fuscus	x					x	x		LC
Olive Python	Liasis olivaceus olivaceus	x					x	x		LC
Pilbara Olive Python	Liasis olivaceus barroni						x			LC
Carpet and Green Pythons, Genus *Morelia*										
Centralian Carpet Python	Morelia bredli							x		LC
Rough-scaled Python	Morelia carinata						x			LC
Western Carpet Python	Morelia imbricata					x	x			LC
Carpet Python	Morelia spilota spilota	x	x	x			x	x		LC
Inland Carpet Python	Morelia spilota metcalfei	x	x	x		x				LC
Green Python	Morelia viridis	x								LC
Scrub and Oenpelli Pythons Genus *Simalia*										
Scrub Python	Simalia amethistina	x								LC
Oenpelli Python	Simalia oenpelliensis							x		VU
File Snakes (Acrochordidae)										
File Snakes, Genus *Acrochordus*										
Arafura File Snake	Acrochordus arafurae	x						x		LC
Little File Snake	Acrochordus granulatus	x					x	x		LC
Typical Snakes (Colubridae)										
Cat Snakes, Genus *Boiga*										
Brown Tree Snake	Boiga irregularis	x	x				x	x		LC
Bronzebacks and Tree Snakes, Genus *Dendrelaphis*										
Northern Tree Snake	Dendrelaphis calligastra	x								LC

Common English Name	Scientific Name	Qld	NSW	Vic	Tas	SA	WA	NT	Territories	IUCN
Common Tree Snake	*Dendrelaphis punctulata*	x	x				x	x		LC
Wolf Snakes, Genus *Lycodon*										
Common Wolf Snake	*Lycodon capucinus*								x	LC
Rat Snakes, Genus *Pantherophis*										
Corn Snake	*Pantherophis guttatus*	x	x	x						LC
Ground Snakes, Genus *Stegonotus*										
Australian Slaty-grey Snake	*Stegonotus australis*	x						x		LC
Water Snakes (Natricidae)										
Keelbacks, Genus *Tropidonophis*										
Keelback	*Tropidonophis mairii*	x	x				x	x		LC
Cobras, Sea Kraits and Sea Snakes (Elapidae)										
Death Adders, Genus *Acanthophis*										
Common Death Adder	*Acanthophis antarcticus*	x	x	x		x	x			VU
Barkly Tableland Death Adder	*Acanthophis hawkei*	x						x		VU
North-western Death Adder	*Acanthophis lancasteri*						x			DD
Northern Death Adder	*Acanthophis praelongus*	x								LC
Desert Death Adder	*Acanthophis pyrrhus*	x				x	x	x		LC
Papuan Death Adder	*Acanthophis rugosus*	x					x	x		LC
Pilbara Death Adder	*Acanthophis wellsei*						x			LC
Sea Snakes, Genus *Aipysurus*										
Short-nosed Sea Snake	*Aipysurus apraefrontalis*						x			CE
Dubois' Sea Snake	*Aipysurus duboisii*	x					x	x		LC
Stagger-banded Sea Snake	*Aipysurus eydouxii*	?					x			LC
Leaf-scaled Sea Snake	*Aipysurus foliosquama*						x			CE
Dusky Sea Snake	*Aipysurus fuscus*						x			EN
Olive Sea Snake	*Aipysurus laevis*	x					x	x		LC
Mosaic Sea Snake	*Aipysurus mosaicus*	x					x	x		LC
Shark Bay Sea Snake	*Aipysurus pooleorum*						x			VU
Mjoberg's Sea Snake	*Aipysurus tenuis*						x			VU
Warros, Genus *Antaioserpens*										
North-eastern Plain-nosed Burrowing Snake	*Antaioserpens albiceps*	x								LC
Robust Burrowing Snake	*Antaioserpens warro*	x								DD
Copperheads, Genus *Austrelaps*										
Pygmy Copperhead	*Austrelaps labialis*					x				VU
Highland Copperhead	*Austrelaps ramsayi*		x	x						LC
Lowland Copperhead	*Austrelaps superbus*		x	x	x	x				LC
Shovel-nosed Snakes, Genus *Brachyurophis*										
North-western Shovel-nosed Snake	*Brachyurophis approximans*						x			LC
Australian Coral Snake	*Brachyurophis australis*	x	x	x		x				LC
Einasleigh Shovel-nosed Snake	*Brachyurophis campbelli*	x								LC
Western Narrow-banded Shovel-nosed Snake	*Brachyurophis fasciolatus fasciolatus*						x			LC
Eastern Narrow-banded Shovel-nosed Snake	*Brachyurophis fasciolatus fasciata*	x	x			x	x	x		LC
Unbanded Shovel-nosed Snake	*Brachyurophis incinctus*	x						x		LC
Arnhem Shovel-nosed Snake	*Brachyurophis morrisi*							x		LC
Northern Shovel-nosed Snake	*Brachyurophis roperi*	x					x	x		LC
Southern Shovel-nosed Snake	*Brachyurophis semifasciatus*	x				x	x	x		LC
Crowned Snakes, Genus *Cacophis*										
Northern Crowned Snake	*Cacophis churchilli*	x								LC
White-crowned Snake	*Cacophis harriettae*	x	x							LC
Southern Dwarf Crowned Snake	*Cacophis krefftii*	x	x							LC

Common English Name	Scientific Name	Qld	NSW	Vic	Tas	SA	WA	NT	Territories	IUCN
Golden-crowned Snake	Cacophis squamulosus	x	x							LC
Small-eyed Snakes, Genus Cryptophis										
Carpentaria Snake	Cryptophis boschmai	x								LC
Pink Snake	Cryptophis incredibilis	x								LC
Eastern Small-eyed Snake	Cryptophis nigrescens	x	x	x						LC
Black-striped Snake	Cryptophis nigrostriatus	x								LC
Northern Small-eyed Snake	Cryptophis pallidiceps						x	x		LC
Whip Snakes, Genus Demansia										
Narrow-headed Whip Snake	Demansia angusticeps						x	x		LC
Black-necked Whip Snake	Demansia calodera						x			LC
Carpentarian Whip Snake	Demansia flagellatio	x								LC
Olive Whip Snake	Demansia olivacea						x	x		LC
Greater Black Whip Snake	Demansia papuensis	x					x	x		LC
Yellow-faced Whip Snake	Demansia psammophis	x	x	x		x				LC
Sombre Whip Snake	Demansia quaesitor	x					x	x		LC
Reticulated Whip Snake	Demansia reticulata reticulata						x			LC
Centralian Whip Snake	Demansia reticulata cupriceps	x	x			x	x	x		LC
Crack-dwelling Whip Snake	Demansia rimicola	x	x			x	x	x		LC
Rufous Whip Snake	Demansia rufescens						x			LC
Shine's Whip Snake	Demansia shinei						x	x		LC
Grey Whip Snake	Demansia simplex						x	x		LC
Collared Whip Snake	Demansia torquata	x								LC
Lesser Black Whip Snake	Demansia vestigiata	x					x	x		LC
Mud Snakes, Genus Denisonia										
De Vis' Banded Snake	Denisonia devisi	x	x	x						LC
Ornamental Snake	Denisonia maculata	x								DD
White-lipped Snakes, Genus Drysdalia										
White-lipped Snake	Drysdalia coronoides		x	x	x	x				LC
Masters' Snake	Drysdalia mastersii			x		x	x			LC
Mustard-bellied Snake	Drysdalia rhodogaster		x							LC
Bardicks, Genus Echiopsis										
Bardick	Echiopsis curta		x	x		x	x			LC
Snakes, Genus Elapognathus										
Western Crowned Snake	Elapognathus coronatus						x			LC
Short-nosed Snake	Elapognathus minor						x			NT
Turtle-headed Sea Snake, Genus Emydocephalus										
Turtle-headed Sea Snake	Emydocephalus annulatus	x	x				x	x		LC
Mangrove Sea Snake, Genus Ephalophis										
Mangrove Sea Snake	Ephalophis greyae						x			LC
Naped Snakes, Genus Furina										
Yellow-naped Snake	Furina barnardi	x								LC
Red-naped Snake	Furina diadema	x	x	x		x				LC
Orange-naped Snake	Furina ornata	x					x	x	x	LC
Snakes, Genus Glyphodon										
Dunmall's Snake	Glyphodon dunmalli	x	x							DD
Brown-headed Snake	Glyphodon tristis	x							x	LC
Swamp Snakes, Genus Hemiaspis										
Grey Snake	Hemiaspis damelii	x	x							E
Marsh Snake	Hemiaspis signata	x	x							LC
Broad-headed Snakes, Genus Hoplocephalus										
Pale-headed Snake	Hoplocephalus bitorquatus	x	x							LC
Broad-headed Snake	Hoplocephalus bungaroides		x							VU
Stephens' Banded Snake	Hoplocephalus stephensii	x	x							NT
Black-ringed Mangrove Snake, Genus Hydrelaps										

Common English Name	Scientific Name	Qld	NSW	Vic	Tas	SA	WA	NT	Territories	IUCN
Black-ringed Mangrove Snake	Hydrelaps darwiniensis	x					x	x		LC
Typical Sea Snakes, Genus Hydrophis										
Black-headed Sea Snake	Hydrophis atriceps	x						x		LC
Belcher's Sea Snake	Hydrophis belcheri	x						x		LC
Dwarf Sea Snake	Hydrophis caerulescens	x								LC
Slender-necked Sea Snake	Hydrophis coggeri	x					x	x		LC
Spine-bellied Sea Snake	Hydrophis curta	x	x				x	x		LC
Geometrical Sea Snake	Hydrophis czeblukovi						x			LC
Rough-scaled Sea Snake	Hydrophis donaldi	x								LC
Elegant Sea Snake	Hydrophis elegans	x	x				x	x		LC
Plain Sea Snake	Hydrophis inornatus	?					?	?		LC
Spectacled Sea Snake	Hydrophis kingii	x					x	x		LC
Laboute's Sea Snake	Hydrophis laboutei	?								DD
Small-headed Sea Snake	Hydrophis macdowelli	x	x				x	x		LC
Olive-headed Sea Snake	Hydrophis major	x	x				x	x		LC
Black-banded Robust Sea Snake	Hydrophis melanosoma	?								LC
Spotted Sea Snake	Hydrophis ornatus	x	x	x	x		x	x		LC
Large-headed Sea Snake	Hydrophis pacificus	x						x		LC
Horned Sea Snake	Hydrophis peronii	x					x	x		LC
Yellow-bellied Sea Snake	Hydrophis platurus	x	x	x	x		x	x		LC
Stokes' Sea Snake	Hydrophis stokesii	x	x				x	x		LC
Plain-banded Sea Snake	Hydrophis vorisi	?								LC
Australian Beaked Sea Snake	Hydrophis zweifeli	x	x				x	x		LC
Sea Kraits, Genus Laticauda										
White-lipped Sea Krait	Laticauda colubrina	x					x	x		LC
Black-lipped Sea Krait	Laticauda laticaudata	x						x		LC
Slender Sea Snake, Genus Microcephalophis										
Slender Sea Snake	Microcephalophis gracilis	x						x		LC
Snakes, Genus Neelaps										
Black-naped Burrowing Snake	Neelaps bimaculatus					x	x			LC
Black-striped Burrowing Snake	Neelaps calonotus						x			NT
Tiger Snakes, Genus Notechis										
Common Tiger Snake	Notechis scutatus scutatus	x	x	x		x				LC
Krefft's Tiger Snake	Notechis scutatus ater					x				VU
Tasmanian Tiger Snake	Notechis scutatus humphreysi				x					LC
Peninsula Tiger Snake	Notechis scutatus niger					x				LC
Western Tiger Snake	Notechis scutatus occidentalis						x			LC
Chappell Island Tiger Snake	Notechis scutatus serventyi				x					VU
Taipans, Genus Oxyuranus										
Inland Taipan	Oxyuranus microlepidotus	x	x	x		x		x		LC
Coastal Taipan	Oxyuranus scutellatus scutellatus	x					x	x		LC
Papuan Taipan	Oxyuranus scutellatus canni	x								LC
Western Desert Taipan	Oxyuranus temporalis						x	x		LC
Northern Mangrove Snake, Genus Parahydrophis										
Northern Mangrove Snake	Parahydrophis mertoni	x						x		LC
Hooded Snakes, Genus Parasuta										
Dwyer's Snake	Parasuta dwyeri	x	x	x						LC
Little Whip Snake	Parasuta flagellum		x	x		x				LC
Gould's Hooded Snake	Parasuta gouldii						x			LC
Monk Snake	Parasuta monachus					x	x	x		LC
Mitchell's Short-tailed Snake	Parasuta nigriceps		x	x		x	x			LC
Mallee Black-headed Snake	Parasuta spectabilis spectabilis		x	x		x				LC
Bush's Black-headed Snake	Parasuta spectabilis bushi						x			DD

Common English Name	Scientific Name	Qld	NSW	Vic	Tas	SA	WA	NT	Territories	IUCN
Nullabor Black-headed Snake	*Parasuta spectabilis nullarbor*					x	x			DD
Lake Cronin Snake, Genus *Paroplocephalus*										
Lake Cronin Snake	*Paroplocephalus atriceps*						x			LC
Black Snakes, Genus *Pseudechis*										
Mulga Snake	*Pseudechis australis*	x	x			x	x	x		LC
Spotted Mulga Snake	*Pseudechis butleri*						x			LC
Collett's Snake	*Pseudechis colletti*	x								LC
Spotted Black Snake	*Pseudechis guttatus*	x	x							LC
Eastern Pygmy Mulga Snake	*Pseudechis pailsei*	x								LC
Papuan Black Snake	*Pseudechis papuanus*	x								DD
Red-bellied Black Snake	*Pseudechis porphyriacus*	x	x	x		x				LC
Western Pygmy Mulga Snake	*Pseudechis weigeli*						x	x		LC
Brown Snakes, Genus *Pseudonaja*										
Dugite	*Pseudonaja affinis affinis*					x	x			LC
Rottnest Island Dugite	*Pseudonaja affinis exilis*						x			LC
Tanner's Brown Snake	*Pseudonaja affinis tanneri*						x			LC
Shield-snouted Brown Snake	*Pseudonaja aspidorhyncha*	x	x	x		x		x		LC
Speckled Brown Snake	*Pseudonaja guttata*	x					x	x		LC
Peninsula Brown Snake	*Pseudonaja inframacula*					x	x			LC
Ingram's Brown Snake	*Pseudonaja ingrami*	x						x		LC
Western Brown Snake	*Pseudonaja mengdeni*	x	x			x	x	x		LC
Ringed Brown Snake	*Pseudonaja modesta*	x	x			x	x	x		LC
Northern Brown Snake	*Pseudonaja nuchalis*	x					x	x		LC
Eastern Brown Snake	*Pseudonaja textilis*	x	x	x		x	x	x		LC
Square-nosed Snake, Genus *Rhinoplocephalus*										
Square-nosed Snake	*Rhinoplocephalus bicolor*						x			LC
Burrowing Snakes, Genus *Simoselaps*										
Desert Burrowing Snake	*Simoselaps anomalus*					x	x	x		LC
Jan's Burrowing Snake	*Simoselaps bertholdi*					x	x	x		LC
West Coast Burrowing Snake	*Simoselaps littoralis*						x			LC
Dampierland Burrowing Snake	*Simoselaps minimus*						x			LC
Curl Snakes, Genus *Suta*										
Rosen's Snake	*Suta fasciata*						x			LC
Ord Curl Snake	*Suta ordensis*						x	x		DD
Little Spotted Snake	*Suta punctata*	x					x	x		LC
Curl Snake	*Suta suta*	x	x	x		x	x	x		LC
Rough-scaled Snake, Genus *Tropidechis*										
Rough-scaled Snake	*Tropidechis carinatus*	x	x							LC
Bandy-bandys, Genus *Vermicella*										
Bandy-bandy	*Vermicella annulata*	x	x	x		x				LC
Intermediate Bandy-bandy	*Vermicella intermedia*							x		LC
Narrow-banded Bandy-bandy	*Vermicella multifasciata*						x	x		LC
Cape York Bandy-bandy	*Vermicella parscauda*	x								NE
Pilbara Bandy-bandy	*Vermicella snelli*						x			LC
Centralian Bandy-bandy	*Vermicella vermiformis*	x						x		LC
Rear-fanged Mud Snakes (Homalopsidae)										
Bockadams, Genus *Cerberus*										
Australian Bockadam	*Cerberus australis*	x					x	x		LC
Crab-eating Snakes, Genus *Fordonia*										
White-bellied Mangrove Snake	*Fordonia leucobalia*	x					x	x		LC
Mangrove Snakes, Genus *Myron*										
Roebuck Bay Mangrove Snake	*Myron resetari*						x			LC
Richardson's Mangrove Snake	*Myron richardsonii*	x					x	x		LC
Macleay's Water Snake, Genus *Pseudoferania*										

Common English Name	Scientific Name	Qld	NSW	Vic	Tas	SA	WA	NT	Territories	IUCN
Macleay's Water Snake	Pseudoferania polylepis	x						x		LC
Blind Snakes (Typhlopidae)										
Australian Blind Snakes, Genus Anilios										
Small-headed Blind Snake	Anilios affinis	x	x							LC
Sand-diving Blind Snake	Anilios ammodytes						x			LC
No-spined Blind Snake	Anilios aspina	x								LC
Southern Blind Snake	Anilios australis						x			LC
Shovel-nosed Blind Snake	Anilios batillus		x							DD
Dark-spined Blind Snake	Anilios bicolor		x	x		x	x			LC
Prong-snouted Blind Snake	Anilios bituberculatus	x	x	x		x	x	x		LC
Faint-striped Blind Snake	Anilios broomi	x								LC
Centralian Blind Snake	Anilios centralis							x		LC
Cape York Striped Blind Snake	Anilios chamodracaena	x								LC
Northern Blind Snake	Anilios diversus	x					x	x		LC
Interior Blind Snake	Anilios endoterus	x	x			x	x	x		LC
Miner Blind Snake	Anilios fossor							x		LC
Gane's Blind Snake	Anilios ganei						x			LC
Long-beaked Blind Snake	Anilios grypus	x					x	x		LC
Top End Blind Snake	Anilios guentheri						x	x		LC
Pale-headed Blind Snake	Anilios hamatus						x			LC
Kimberley Deep-soil Blind Snake	Anilios howi						x			DD
Fassifern Blind Snake	Anilios insperatus	x								CE
Kimberley Shallow-soil Blind Snake	Anilios kimberleyensis						x	x		LC
Murchison Blind Snake	Anilios leptosoma						x			LC
Cape York Blind Snake	Anilios leucoproctus	x								LC
Robust Blind Snake	Anilios ligatus	x	x			x	x	x		LC
Barrow Island Blind Snake	Anilios longissimus						x			DD
Buff-snouted Blind Snake	Anilios margaretae						x			LC
Small-eyed Blind Snake	Anilios micromma						x			DD
Groote Eylandt Dwarf Blind Snake	Anilios minimus							x		LC
Thread-like Blind Snake	Anilios nema							x		LC
Blackish Blind Snake	Anilios nigrescens	x	x	x						LC
Blunt-snouted Blind Snake	Anilios obtusifrons						x			NE
Pilbara Blind Snake	Anilios pilbarensis						x			LC
Rotund Blind Snake	Anilios pinguis						x			LC
Proximus Blind Snake	Anilios proximus	x	x	x		x				LC
Roberts' Blind Snake	Anilios robertsi	x								DD
Cooloola Blind Snake	Anilios silvia	x								LC
Splendid Blind Snake	Anilios splendidus						x			DD
Sharp-snouted Blind Snake	Anilios systenos						x			NE
North-eastern Blind Snake	Anilios torresianus	x								LC
Darwin Blind Snake	Anilios tovelli							x		LC
Sandamara Blind Snake	Anilios troglodytes						x			LC
Claw-snouted Blind Snake	Anilios unguirostris	x					x	x		LC
Beaked Blind Snake	Anilios waitii						x			LC
Brown-snouted Blind Snake	Anilios wiedii	x	x							LC
Yampi Blind Snake	Anilios yampiensis						x			DD
Yirrkala Blind Snake	Anilios yirrikalae							x		LC
West Kimberley Blind Snake	Anilios zonula						x			DD
Indonesian Blindsnakes, Genus Indotyphlops										
Flowerpot Snake	Indotyphlops braminus	x					x	x		LC
Long-tailed Blind Snakes, Genus Ramphotyphlops										
Christmas Island Blind Snake	Ramphotyphlops exocoeti								x	EN

▪ FURTHER READING ▪

WEBSITES

Atlas of Living Australia www.ala.org.au
Australian Faunal Directory www.biodiversity.org.au
Australian Reptile Online Database (AROD) www.arod.com.au/arod
Australian Venom Research Unit
www.biomedicalsciences.unimelb.edu.au/departments/pharmacology/engage/avru
Australia's Wildlife www.australiaswildlife.com
Herpmapper www.herpmapper.org
Nature 4 You www.wildlifedemonstrations.com

REFERENCES

Aplin, K. P. (1998) Three new blindsnakes (Squamata: Typhlopidae) from northwestern Australia. *Records of the Western Australian Museum* 19, 1–12.

Barker, D. G. & Barker, T. M. (1994) *Pythons of the World. Vol. 1, Australia.* The Herpetocultural Library, Advanced Vivarium Systems Inc. Lakeside, Calif.

Cogger, H. (2018) *Reptiles and Amphibians of Australia* (7th edn). CSIRO Publishing, Collingwood.

Couper, P. J., Covacevich, J. A. & Wilson, S. K. (1998) Two new species of *Ramphotyphlops* (Squamata:Typhlopidae) from Queensland. *Memoirs of the Queensland Museum* 42 (2) 459–464.

Couper, P. J., Peck, S. R., Emery, J-P., Keogh, J. S. (2016) Two snakes from eastern Australia (Serpentes: Elapidae); a revised concept of *Antaioserpens warro* (De Vis) and a redescription of *A. albiceps* (Boulenger). *Zootaxa* 4097 (3) 227–231.

Derez, C. M., Arbuckle, K., Ruan, Z., Xie, B., Huang, Y., Dibben, L., Shi, Q., Vonk, F. & Fry, B. G. (2018) A new species of bandy-bandy (*Vermicella*: Serpentes: Elapidae) from the Weipa region, Cape York, Australia. *Zootaxa* 4446 (1) 1–12.

Ehmann, H. (1992) *Encyclopaedia of Australian Animals – Reptiles.* Collins Angus & Robertson, Sydney, NSW.

Eipper, S. C. (2012) *A Guide to Australian Snakes in Captivity: Elapids and Colubrids.* Reptile Keeper Publications: Tweed Heads, NSW.

Elliott, A. (2012) *A Guide to Australian Snakes in Captivity: Pythons.* Reptile Keeper Publications: Tweed Heads, NSW.

Ellis, R. J. (2016) A new species of blindsnake (Scolecophidia: Typhlopidae: *Anilios*) from the Kimberley region of Western Australia. *Herpetologica* 72 (3), 271–278.

Ellis, R. J., Doughty, P., Donnellan, S. C., Marin, J. & Vidal, N. (2017) Worms in the sand: systematic revision of the Australian blindsnake *Anilios leptosoma* (Robb, 1972) species complex (Squamata: Scolecophidia: Typhlopidae) from the Geraldton sandplain, with description of two new species. *Zootaxa* 4323(1) 1–24.

Gow, G. F. (1989) *A Complete Guide to Australian Snakes.* Angus and Robertson, Sydney, NSW, Australia.

Greer, A. E. (1997) *The Biology and Evolution of Australian Snakes.* Surrey Beatty and Sons, Chipping Norton, NSW, Australia.

Günther, A. (1864) *The Reptiles of British India. London* (Taylor & Francis), xxvii + 452 pp.

Heatwole, H. (1999) *Sea Snakes, Australian Natural History Series.* University of NSW Press, Sydney, NSW.

Horner, P. (1998) *Simoselaps morrisi* sp.nov (Elapidae), a species of snake from the Northern Territory. *The Beagle. Records of the Museums and Art Galleries of the Northern Territory* 14, 63–70.

Ingram, G. J. & Covacevich, J. A. (1993) Two new species of striped blindsnakes. *Memoirs of the Queensland Museum* 34 (1) 181–184.

Kaiser, C. M., Kaiser, H. & O'Shea, M. (2018) The taxonomic history of Indo-Papuan groundsnakes, genus *Stegonotus* Duméril et al.,1854 (Colubridae), with some taxonomic revisions and the designation of a neotype for *S. parvus* (Meyer, 1874). *Zootaxa* 4512 1–73.

Kuch, U. & Yuwono, F. B. (2002) First record of Brown Snakes *Pseudonaja* cf. *textilis* (Duméril, Bibron & Duméril, 1854) from Papua, Indonesia. *HERPETOZOA* 15(1/2), 75–78.

Lincoln, R. J., Boxshall, G. A. & Clark, P. F. (1982) *A Dictionary of Ecology, Evolution and Systematics*. Cambridge University Press, Cambridge.

McDowell, S. B. (1972) The species of *Stegonotus* (Serpentes: Colubridae) in Papua New Guinea. *Zoologische Mededelingen* 47(2), 6–26.

Mirtschin, P. J., Rasmussen, A. R. & Weinstein, S. A. (2017) *Australia's Dangerous Snakes, Identification, Biology and Envenoming*. CSIRO Publishing, Clayton South, Victoria.

Murphy, J. C. (2007) *Homalopsid Snakes – Evolution in the Mud*. Krieger Publishing, Florida, USA.

Murphy, J. C. (2011) The nomenclature and systematics of some Australasian homalopsid snakes (Squamata: Serpentes: Homalopsidae). *The Raffles Bulletin of Zoology* 59 (2), 229–236.

Ruane, S., Richards, S. J., McVay, J., Tjaturadi, B., Krey, K. & Austin, C. C. (2017) Cryptic and non-cryptic diversity in New Guinea groundsnakes of the genus *Stegonotus*, Duméril, Bibron and Duméril, 1854: a description of four new species (Squamata: Colubridae), *Journal of Natural History*.

Rasmussen, A. R. & Ineich, I. (2000) Sea Snakes of New Caledonia and surrounding waters (Serpentes: Elapidae): first report on the occurrence of *Lapemis curtus* and a description of a new species from the genus *Hydrophis*. *Hamadryad* 25, 91–99.

Rasmussen, A. R., Sanders, K. L., Guinea, M. l. & Amey, A. P. (2014) Sea Snakes in Australian waters (Serpentes: Subfamilies Hydrophiinae and Laticaudinae) – a review with an updated key. *Zootaxa* 3869 (4) 351–371.

Rawlings, L. H., Rabosky, D. L., Donnellan, S. C. & Hutchinson, M. N. (2008) Python phylogenetics: inference from morphology and mitochondrial DNA. *Biological Journal of the Linnean* Society 93 (3) 603–619.

Rowland, P. & Eipper, S. C. (2018) *A Naturalist's Guide to the Dangerous Creatures of Australia*. Beaufoy Publishing, Oxford.

Sanders, K. L., Rasmussen, A. R., Elmberg, J., Mumpuni, Guinea, M., Blias, P., Lee, M. S. Y. & Fry, B. G. (2012) *Aipysurus mosaicus* a new species of egg-eating sea snake (Elapidae, Hydrophiinae) with a redescription of *Aipysurus eydouxii* (Gray 1849). *Zootaxa* 3431, 1–18.

Sanders, K. L., Lee, M. S. Y, Mumpuni, Bertozi, T. & Rasmussen, A. R. (2013) Multilocus phylogeny and recent rapid radiation of the viviparous sea snakes (Elapidae, Hydrophiinae). *Molecular Genetics and Evolution* 66: 575–591.

Scanlon, J. D. (2003) The Australian elapid genus *Cacophis*: morphology and phylogeny of rainforest crowned snakes. *Herpetological Journal* 13:1–20.

Shea, G. & Horner, P. (1997) A new species of *Ramphotyphlops* (Squamata: Typhlopidae) from the Darwin area, with notes on two similar species from Northern Australia. *The Beagle. Records of the Museums and Art Galleries of the Northern Territory* 13, 53–60.

Shea, G. (2015) A new species of *Anilios* (Scolecophidia: Typhlopidae) from Central Australia. *Zootaxa* 4033 (1), 103–116.

Shine, R. (1991) *Australian Snakes – a Natural History*. Reed Books, Balgowlah, New South Wales.

Somaweera, R. (2017) *A Naturalist's Guide to the Reptiles and Amphibians of Bali*. Beaufoy

Publishing, Oxford.

Storr, G. M. (1981) The genus *Ramphotyphlops* (Serpentes: Typhlopidae) in Western Australia. *Records of the Western Australian Museum* 9(3), 235–271.

Storr, G. M. (1984) A new *Ramphotyphlops* (Serpentes: Typhlopidae) from central Australia. *Records of the Western Australian Museum* 11(3), 287–290.

Storr, G. M. (1989) A new *Rhinoplocephalus* (Serpentes: Elapidae) from Western Australia. *Records of the Western Australian Museum* 14 (1), 137–138.

Storr, G. M. (1989) A new *Pseudonaja* (Serpentes: Elapidae) from Western Australia. *Records of the Western Australian Museum* 14 (3), 421–423.

Storr, G. M., Smith, L. A. & Johnstone, R. E. (2002) *Snakes of Western Australia*. Western Australian Museum, Perth.

Sutherland, S. K. & Tibballs, J. (2001) *Australian Animal Toxins* (2nd edn). Oxford University Press, Melbourne.

Swan, G., Shea, G. & Sadlier, R. (2017) *A Field Guide to the Reptiles of NSW*. Reed New Holland, Sydney, NSW.

The IUCN Red List of Threatened Species. www.iucnredlist.org. Downloaded on 20 March 2018.

Venchi, A., Wilson, S. K. & Borsboom, A. C. (2015) A new blindsnake (Serpentes: Typhlopidae) from an endangered habitat in south-eastern Queensland, Australia. *Zootaxa* 3990 (2), 272–278.

Wells, R. & Wellington, C. R. (1985) A classification of the Amphibia and Reptilia of Australia. *Australian Journal of Herpetology* (Supplementary Series) No 1: 1–64.

White, J. (2013) *A Clinician's Guide to Australian Venomous Bites and Stings*. BioCSL, Parkville, Melbourne.

Wilson, S. K. & Swan, G. (2017) *A Complete Guide to Reptiles of Australia* (5th edn). New Holland, Chatswood, Sydney.

Worrell, E. (1961) A new insular Brown Snake. *Proceedings of the Royal Zoological Society of New South Wales for 1958–59*: 56–58.

Zozaya, S. & Macdonald, S. (2017) Snakes of Australia. An electronic field guide to the snakes of Australia. App Version 1.1 (2061).

■ Acknowledgements ■

We, the authors, firstly thank our sons Bailey and Cody, for their wonderful and unconditional support and assistance during the writing of this book.

Special thanks go to the following people for their assistance in making this book possible. This includes, but is not limited to, access to specimens in their care, assistance in the field, supply of images and thought-provoking conversations. Thank you Luke Allen, Allen Allison, Andrew Amey, Cameron Baker, Scott van Barneveld, Shane Black, Daniel Bromley, Clay Bryce, Elliott Budd, Brian Bush, John Cann, Casey Cannon, Tom Charlton, Nathan Clout, Hal Cogger, Patrick Couper, Mark Cowan, Jordan De Jong, Mike Donovan, Paul Doughty, Nathan Dunstan, Graham Edgar, Cody Eipper, Euan Edwards, Adam Elliott, Ryan Ellis, Jon Paul Emery, Jules Farquhar, Brad Foy, Ryan Francis, Bryan Fry, Glen Gaikhorst, Prathamesh Ghadekar, Richie Gilbert, Jamie Gover, Philip Griffin, John Harris, Aaron Hillier, Harry Hines, Paul Horner, David Hunter, Ivan Ineich, Max Jackson, Hinrich Kaiser, Scott Kickham, Gavin Lawrence, Jacob Loyacano, Mike Lyons, Phil Mangion, Brad Maryan, Ross McGibbon, Jake Meney, Peter Mirtschin, the late Michael Moore, Brock Morris, James Nankivell, Angus McNab, Courtenay Noble, Mark O'Shea, Fred Parker, Jarrad Prangell, Dean Purcell, Arne Rasmussen, David Robinson, Peter Rowland, Jodi Rowley, Daniel Rumsey, Mark Sanders, Gunther Schmida, Shawn Scott, Glenn Shea, Ruchira Somaweera, Nigel Sowter, Gary Stephenson, Peter Street, Jason Sulda, Matt Summerville, Gerry Swan, Michael Swan, Steve Swanson, Janne Torkkola, Brad Traynor, Steve Tuckey, Kanishka Ukuwela, Eric Vanderduys, Freek Vonk, Lauren Vonnahme, Harold Voris, Jordan Vos, Richard Wells, Steve Wilson, Justin Wright, Wolfgang Wüster and Anders Zimny. We thank the Australian Museum, Queensland Museum and Western Australian Museum for giving us access to examine specimens.

This book has been greatly improved by the constructive comments of Hal Cogger, Adam Elliott, Angus McNab, Janne Torkkola and the two anonymous reviewers. We would also like to thank our parents Lyndon and Gail Kember, and Colin and Christine Eipper for their assistance and support.

Finally, we thank the publishers, including John Beaufoy, Rosemary Wilkinson and their staff, for the opportunity to write this work, series editor Krystyna Mayer for ensuring readability and consistency in the text, and Sally Bird of Calidris Literary Agency for the establishment and management of these valuable relationships.